Reinforcement Learning From Scratch

Uwe Lorenz

Reinforcement Learning From Scratch

Understanding Current
Approaches - with Examples
in Java and Greenfoot

 Springer

Uwe Lorenz
Neckargemünd
Baden-Württemberg
Germany

This book is a translation of the original German edition "Reinforcement Learning" by Lorenz, Uwe, published by Springer-Verlag GmbH, DE in 2020. The translation was done with the help of artificial intelligence (machine translation by the service DeepL.com). A subsequent human revision was done primarily in terms of content, so that the book will read stylistically differently from a conventional translation. Springer Nature works continuously to further the development of tools for the production of books and on the related technologies to support the authors.

ISBN 978-3-031-09032-5   ISBN 978-3-031-09030-1   (eBook)
https://doi.org/10.1007/978-3-031-09030-1

This Springer imprint is published by the registered company Springer Nature Switzerland AG
The registered company address is: Gewerbestrasse 11, 6330 Cham, Switzerland

Preface

You know, I couldn't do it. I couldn't reduce it to the freshman level. That means we really don't understand it.

R. P. Feynman

Regarding the quotation above, perhaps the content in this book is not quite reduced to a beginner's level, but it should be very suitable, especially for newcomers to machine learning. It is a book for those with some basic knowledge of programming and high school level math. It is useful, for example, for continuing education for teachers and instructors who want to gain insight into programming adaptive agents. The book should also be appropriate for technicians, computer scientists, or programmers who want to better understand RL algorithms by studying and implementing learning these algorithms from scratch themselves (especially if they have been socialized with Java) or for students who want to study machine learning and intelligent agents.

Reinforcement Learning is an area of Machine Learning and belongs to the broad field of Artificial Intelligence. Many believe that reinforcement learning has something to do with artificial neural networks. Although the book also deals with reinforcement learning algorithms that use neural networks, this does not play a central role. Deep reinforcement learning is actually a subset of reinforcement learning, where some functions are implemented with deep neural networks. It may surprise that explanations of the "Asynchronous Advantage Actor-Critic (A3C)" or "Proximal Policy Optimization (PPO)" are possible without the use of neural networks.

We will not dwell on an overly detailed description of mathematical relationships but rather do our own implementations from scratch. Let us use an analogy with a tool such as Google Maps. There are works that are zoomed out very far. They draw an overview that may be completely devoid of mathematics and concrete implementations. This can be useful when you want to give an impression of the landscape. give an impression of the landscape. On the other hand, there are works that get down to the details, some of them want to enable on the use of tools application oriented, quasi enable to plan a concrete route through a real city, others focus mainly on a detailed description of the theory.

My goal is for them to get an understanding of the central processes involved in reinforcement learning. If you want to teach how a gasoline engine works, you do not have to analyze a high-performance model. A rough model showing the movement of the pistons, etc., is sufficient. In addition, there are some essential formulas that deepen the insight, but we will mainly do our own constructions. For me, the best teaching model is the one that teaches the ability to model by yourself. To really understand things, it is necessary to do them yourself. It was also an important motivation for me to reproduce the algorithms in my "native" programming language (Java) in order to understand them better.

An added benefit of this exercise for the reader is that he or she can apply the knowledge to a variety of frameworks, with a good understanding of what each parameter means or what the functions do, and perhaps even work out his or her own custom designs. Therefore, the book is particularly useful if you have been socialized with C or Java style languages and, for example, may want or need to use reinforcement learning libraries in another language such as Python. Java has the image of being somewhat complicated, but there are also some very sophisticated tools for playing with and learning the language. An important and widely used tool developed for teaching object-oriented programming with Java is Greenfoot, which we will use here to program adaptive behavior to some Greenfoot-"actors". We will consider them as interacting agents that are part of (two dimensional) worlds that can created and controlled completely by the learner. I would like to thank Michael Kölling (King's College London) very much, chief designer of Greenfoot, for his comments on this book. I found them insightful, encouraging and motivating.

I would like to dedicate this book to my children Jonas, Elias, Paul and Julia. I promise that now that this project has come to its conclusion, we will do more together again. I would like to thank my wife Anja, without her support this project would not have been possible. I would also like to thank my sister Ulrike aka "HiroNoUnmei" (https://www.patreon.com/hironounmei; April, 2022) for the funny hamster illustrations at the beginning of each chapter.

The accompanying materials (Java programs, explanatory videos, etc.) are accessible via the product page of the book at SpringerLink. Through the site (https://www.facebook.com/ReinforcementLearningJava; April, 2022), it may also be possible to build a small community and initiate an exchange about the contents of the book and the topic of reinforcement learning. Feel free to post substantive contributions on the topic or interesting results. You may also find the opportunity to ask comprehension questions or address points that remain open.

Neckargemünd, Baden-Württemberg, Germany   Uwe Lorenz
2022

Introduction

Understanding grows with active engagement: to 'do' something, to master it, is at the same time to understand it better. Applied to the study of mental processes, this leads to the replication of intelligent behavior with machines.

H. D. Burkhardt[1]

Abstract This introductory section describes what this book is about and for whom it is intended: the topic of "reinforcement learning," as a very exciting subfield of artificial intelligence or machine learning, is to be presented in a form that quickly teaches beginners the most important approaches and some of the central algorithms, and also allows them to conduct their own experiments with it.

It is intended to be a non-fiction book that provides learners or those interested in (or needing to) deal with the content of this area of artificial intelligence with a practical approach to the theory and operation of adaptive agents. However, it is also intended for technicians or teachers who would like to further their education and conduct their own experiments or exercises, also within the framework of their respective courses.

Some "philosophical" issues or criticisms of the field of AI research are briefly reflected upon.

This book may not be without danger, as it is about artificial agents that can learn. At the South by Southwest digital festival in the US state of Texas in 2018, Elon Musk said, "artificial intelligence is much more dangerous than nuclear weapons." Perhaps this is a bit thick; however, it is certainly beneficial for as many people as possible to understand and assess how this technology works. This allows not only to judge what the technology can and should do but also to help shape its further development path. Attempting to contain valuable knowledge would certainly be a futile endeavor in today's world.

The topic of "reinforcement learning" is of particular importance in this context because it involves autonomous machines that, like biological creatures, are

---

[1] Prof. Dr. Hans-Dieter Burkhard was one of my former professors at the Humboldt University of Berlin – in 2004, 2005, and 2008 World Champion in the Four-Legged Robots League at the RoboCup with the "German Team."

presented with problems in a specific environment and optimize their behavior through active processes. However, "reinforcement learning" is also one of the most fascinating areas of machine learning that is often still comparatively little dealt with in German-speaking countries, although spectacular success stories from this field of artificial intelligence repeatedly reach not only the specialist audience but also the broad media public. Some examples: One of the longest studied domains in the history of artificial intelligence is that of chess. Some time ago, it was possible to write programs that could beat human champions, but these used sophisticated, specialized search techniques and hand-crafted evaluation functions as well as enormous databases of archived moves. The "Alpha Zero" series of programs from "Google DeepMind," on the other hand, can achieve superhuman feats in a matter of hours just by learning from games against themselves without any prior human knowledge at all. It ends up significantly outperforming even the best programs to date that used the aforementioned methods. Another spectacular aspect is that the system is not limited to a single game like chess, but is capable of learning all kinds of board games on its own. This includes "Go," probably the most complex board game in many respects, which has been played in Asia for thousands of years and actually requires a kind of "intuitive" interpretation of the situation of the board. Alpha Zero can adapt to a variety of highly complex strategy games extremely successfully and does not require any further human knowledge to do so – the rules of the game alone are sufficient for this. The system thus becomes a kind of universal board game solution system. How did Google DeepMind achieve these successes? Until recently, there was a great deal of agreement that an "intuitive" game with such a large number of states as Go could not be mastered by serial computers in the foreseeable future. In addition, the machine capable of beating human champions in such games is completely devoid of the human knowledge of the games accumulated over millennia.

Similarly, spectacular are the performances within "dynamic" environments, for example, those of "Deep Q Networks" when independently learning arbitrary Atari arcade games or the results in the field of robotics, where systems learn by independent trial and error to successfully execute complex movements such as grasping, running, and jumping, and to master tasks in prepared arenas, such as those posed in the numerous robotics competitions that now exist for all possible levels and ages.

So-called "deep" artificial neural networks with special generalization abilities play a major role here. Astonishing blossoms can be seen, for example, in the automatic transformation of photos into deceptively "real" paintings in the style of a certain artist, where two such "deep" networks interact in such a way that one of them "produces" and another one "criticizes" the products. You can get your own impression of this at https://deepart.io [Feb. 2020].

Recently, elaborate machine learning frameworks like TensorFlow have been made available by relevant players, sometimes for free. Why do companies like Google give such elaborate products to the general public? Arguably, it is also to define standards and create dependencies. Philanthropic generosity is certainly not to be assumed as the primary motive for such corporations. The book also wants to encourage people to take a closer look at the "gears" behind the "frameworks" and to understand them from the ground up.

Concrete implementations of reinforcement learning often seem quite complicated. The "learning processes" depend on many parameters and practical circumstances. They take a lot of computing time, and at the same time, the success is uncertain. However, the ideas of the algorithms behind the learning agents are usually very clear and easy to grasp. Furthermore, we will use live visualizations to observe the learning progress and the currently achieved learning state online.

The topic of "reinforcement learning" is to be presented in a form that quickly teaches beginners the most important approaches and central algorithms, as well as enabling them to conduct their own interesting experiments. Tools such as those used in beginners' courses or in programming classes will be applied. You will also find suggestions for teaching and learning in the book. It is not primarily about operating a high-performance black box, but about understanding, comprehending, evaluating, and perhaps also innovatively developing the algorithms in your own experiments. Driving a car is one thing, understanding how a car's gasoline engine works is another. Although the two are also closely linked: on the one hand, driving also requires certain knowledge of how a car works and, conversely, the designer of a car also determines how the car is driven. In a figurative sense, after some theoretical preliminary considerations, we will each build and try out some "engines" and "soapboxes" ourselves in order to understand how the technology works. In addition, driving with market-ready products will also be made easier. After all, it is by no means the case that the use of ready-made libraries for reinforcement learning works without problems right away.

"Chessboard worlds," so-called gridworlds, like ◉ Fig. 1 play a major role in introductory courses in programming. However, these are not board games but

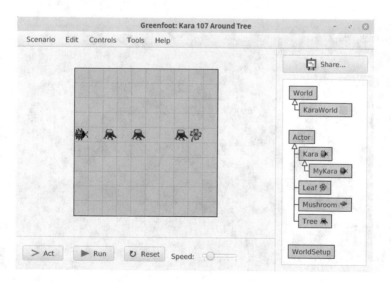

Fig. 1 Gridworld "Kara"

two-dimensional grids in which various objects, that is, "treasures," "traps," "walls," and the like, as well as movable figures are located.

Furthermore, robotic hardware, usually equipped with a differential drive and some simple sensors, such as touch or color sensors, is widely used in teaching, for example, with kits from the toy manufacturer LEGO, as shown in ◉ Fig. 2, Fischertechnik, or open-source projects such as "Makeblock."

With such aids, basic algorithmic structures such as sequences, conditional statements, or repetition loops are taught. However, the functionality of the elementary algorithmic structures is thereby recognized quite quickly. The funny figures in the chessboard worlds, such as the Kara beetle or the Java hamster or also the robot creations, arouse the motivation to program "really" intelligent behavior; however, on the level of the algorithmic basic structures, the behavior of such "robots" usually remains very mechanical and hardly flexible – intelligent behavior looks different.

Interestingly, the standard academic literature on reinforcement learning is also full of such "gridworlds" and simple robots. This is because they offer clear advantages: on the one hand, they are complex enough for interesting experiments and very illustrative, but on the other hand, because of their simplicity, they are easy to understand and allow mathematical penetration. In this introductory textbook and experimental book, these simple "worlds" will first give us a descriptive approach to the algorithms of adaptive agents, and in later chapters we will also deal with "continuous" and dynamic scenarios. The book is intended for learners or interested people who want (or need) to deal with this area of artificial intelligence; furthermore, it is also intended for teachers or technicians who want to further educate themselves and conduct illustrative exercises with their students or their own experiments.

Fig. 2 Robot with differential drive

This book has the peculiarity that the algorithms are not presented in the programming language Python but in the language Java, which is widely used by software developers and also in teaching, especially in connection with object-oriented programming. For most of those who were socialized in the artificial intelligence scene of the early 2000s, Java often still represents a kind of "mother tongue." Recently, big players such as Amazon or LinkedIn have also come onto the market with new, interesting, and free tools for the Java community.

The buzzwords "artificial intelligence" and "machine learning" are currently on everyone's lips. "Artificial intelligence" is the much broader term here. This also includes, for example, rule-based systems of GOFAI ("Good Old-Fashioned Artificial Intelligence"), which have not only produced (old-fashioned) chess programs but also certain voice assistants, chatbots, and the like. In such old-fashioned "expert systems," "knowledge" is symbolically represented and interconnected with production rules, similar to if-then statements. Learning is understood here primarily as an assimilation and processing of symbolic "knowledge." "Machine learning," on the other hand, is currently more associated with the supervised training of pattern recognizers, especially with artificial neural networks. These technologies are of great practical importance and have been used by US companies, in particular, for decades with enormous economic success.

"Reinforcement learning," on the other hand, cannot really be assigned to either of the two fields mentioned. It has a much broader and, above all, more holistic perspective on learning systems, in which, for example, the embedding of the learning system in its environment is taken into account to a much greater extent. Reinforcement learning is essentially about building actively learning, that is, autonomously acting, agents that behave increasingly successfully in their environment, improving through trial and error. The challenge is to build internal structures during the self-acting exploration process that guide the agent behavior in an increasingly purposeful way. This also requires combining insights from different areas of artificial intelligence and machine learning.

This is not surprising, considering that the scenario of reinforcement learning – successful action within an environmental system – corresponds much more to the biological roots of cognition, where all cognitive abilities are applied in combination according to natural requirements and are permanently developed. Currently, AI research in general is taking more notice of the biological origins of the cognitive apparatus than in the past. Concepts such as "situated AI approach" and, in an extended sense, "embodiment," that is, the nature of the "body" and of sensory and motor abilities, are playing a greater role. On the one hand, attempts are being made in the neurosciences, for example, to better understand the functioning of biological cognitive systems or to artificially reproduce and simulate the dynamics of living systems, including experiments with "artificial life."

The interest in "artificial intelligence" started at least already in the mechanical age, i.e. long before Turing's times, for example with the mechanical calculating machines of Leibniz or Pascal. There were also many curious constructions cf. Fig. 3, and many aberrations, but also many important insights were gained, not only from a technical but also from an epistemological point of view.

Fig. 3 Vaucanson's mechanical duck (1738) could flap its wings, quack and drink water. It even had an artificial digestive apparatus: grains that were picked up by it were "digested" in a chemical reaction in an artificial intestine and then excreted in a lifelike consistency. (Source: Wikipedia)

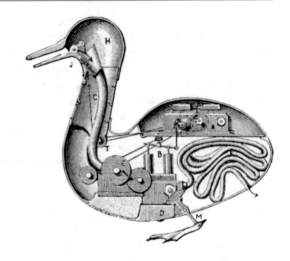

In "reinforcement learning," classical, model-based AI approaches, which have produced, for example, various search algorithms, can productively combine with "situated," "model-free," or "connectionist" approaches, where, for example, neural deep-learning networks then also play a role, as we will see, for example, with the "Alpha Go Zero" algorithm.

The topic of artificial intelligence also arouses fears, not only because of the dangers that the new technology brings with it. Some also have fundamental criticisms, such as Julian Nida-Rümelin, who sees the end of enlightened humanism coming with the advent of a "machine man." Also to mention is Weizenbaum's classic critique *Computer Power and Human Reason*, where he criticizes the simplistic concept of intelligence used by AI researchers and calls the idea of "artificial intelligence" a "perverse, grandiose fantasy." On the other hand, we find "AI prophets" who exaggerate some meager results, mystify and praise their "secret knowledge" in order to adulate themselves and their guild.

The book wants to contribute to a better understanding of different aspects of this technology, to assess the limits, but also the enormous potentials, more realistically and to better judge mystifying statements, but also critical remarks. On the last pages of the book, more philosophical questions and perspectives will be discussed again. As a rule, in my opinion, such critical statements do not touch the foundations of AI research as fundamentally as the authors often suggest. The productive consideration of the criticisms often leads to interesting questions and important further developments.

Contents

1 Reinforcement Learning as a Subfield of Machine Learning 1
1.1 Machine Learning as Automated Processing of Feedback
from the Environment 2
1.2 Machine Learning 3
1.3 Reinforcement Learning with Java 8
Bibliography 13

2 Basic Concepts of Reinforcement Learning 15
2.1 Agents 16
2.2 The Policy of the Agent 18
2.3 Evaluation of States and Actions (Q-Function, Bellman
Equation) 20
Bibliography 22

3 Optimal Decision-Making in a Known Environment 23
3.1 Value Iteration 25
3.1.1 Target-Oriented Condition Assessment ("Backward
Induction") 25
3.1.2 Policy-Based State Valuation (Reward Prediction) 34
3.2 Iterative Policy Search 36
3.2.1 Direct Policy Improvement 37
3.2.2 Mutual Improvement of Policy and Value Function 38
3.3 Optimal Policy in a Board Game Scenario 43
3.4 Summary 46
Bibliography 46

4 Decision-Making and Learning in an Unknown Environment 47
4.1 Exploration vs. Exploitation 49
4.2 Retroactive Processing of Experience ("Model-Free
Reinforcement Learning") 51
4.2.1 Goal-Oriented Learning ("Value-Based") 51
4.2.2 Policy Search 66
4.2.3 Combined Methods (Actor-Critic). 84

4.3 Exploration with Predictive Simulations ("Model-Based
    Reinforcement Learning") 96
    4.3.1 Dyna-Q. 97
    4.3.2 Monte Carlo Rollout 101
    4.3.3 Artificial Curiosity 107
    4.3.4 Monte Carlo Tree Search (MCTS). 111
    4.3.5 Remarks on the Concept of Intelligence 118
4.4 Systematics of the Learning Methods 120
Bibliography 121

5 **Artificial Neural Networks as Estimators for State Values
   and the Action Selection** 123
   5.1 Artificial Neural Networks. 125
       5.1.1 Pattern Recognition with the Perceptron 128
       5.1.2 The Adaptability of Artificial Neural Networks. . 131
       5.1.3 Backpropagation Learning. 146
       5.1.4 Regression with Multilayer Perceptrons 149
   5.2 State Evaluation with Generalizing Approximations. 152
   5.3 Neural Estimators for Action Selection 163
       5.3.1 Policy Gradient with Neural Networks 163
       5.3.2 Proximal Policy Optimization 165
       5.3.3 Evolutionary Strategy with a Neural Policy 169
   Bibliography 173

6 **Guiding Ideas in Artificial Intelligence over Time** 175
   6.1 Changing Guiding Ideas 176
   6.2 On the Relationship Between Humans and Artificial
       Intelligence. 181
   Bibliography 184

Reinforcement Learning as a Subfield of Machine Learning

1

In evolution, being smart counts for nothing if it doesn't lead to acting smart.

Michael Tomasello (2014)

Abstract

This chapter deals with a behavior-oriented concept of machine learning and the classification of reinforcement learning in the field of machine learning in general. A rough overview of the different principles of machine learning is given, and it is explained how they differ from the technical approach. Subsequently, the special features of the implementation of reinforcement learning algorithms with the Java programming language are discussed.

1.1   Machine Learning as Automated Processing of Feedback from the Environment

A great deal of media interest has been attracted by the complex adaptability of artificial neural networks (ANNs). "Deep"-ANNs can, for example, learn to distinguish images of dogs and cats. This sounds trivial, but for decades it represented an almost unsolvable task with the hardware and software available up to now. With the technical solution of this pattern recognition problem, many completely new application possibilities for computer systems were created, for example, in medical diagnostics, industrial production, scientific evaluation of data, marketing, finance, military or security technology, and many more. These innovations are tremendous, and in Star Trek parlance, it is most amazing and fascinating that we are living in an age where things are being done and works are being created on a grand scale that has never been done before in the tens of thousands of years of human history.

However, pattern recognition is only one part of machine learning, namely, the so-called "supervised learning," especially the part that works with distributed internal representations. Although the training of artificial neural networks will be discussed later, this book will not deal with pattern recognition.

2011 Turing Award winner Judea Pearl said, "Every impressive achievement of 'deep learning' boils down to fitting a curve to data. From a mathematical point of view, it doesn't matter how cleverly you do it—it remains curve fitting, albeit complex and far from trivial."[1]

Function fits to a data set in this sense represent only one aspect of system behavior, what we would commonly call "intelligent." With the help of a curve well fitted to a given input data set, we can indeed interpolate the function values, e.g., "cat" or "dog," for previously never seen high-dimensional "arguments," but intelligent system behavior clearly goes beyond this. In particular, if we want to speak of adaptive artificial intelligence, we would like to include activities such as the sensible control of a vacuum cleaner, the opening of a door by a robot arm, or competent recommendations for action, e.g., on the stock exchange, or even exciting opponents in board games such as chess and go, or in gaming in general.

Here, AI software must not only evaluate diverse, partially interdependent states but must also act in a far-sighted manner. The behavior of a pattern classifier is actually limited to classification into specific categories. Training of such a system is done by immediate feedback of a knowing rather authoritarian teacher, "Yes—done right." or "No—done wrong. That is the correct answer. Please improve." This is not sufficient for scenarios such as those mentioned above; here we receive no feedback at all most of the time about whether we are on a goal-directed path or whether we would have done better differently at some point. Moreover, at the end of an episode, we sometimes do not even find any knowing feedback about what would have been the right action, but we only collect more or less large rewards or "punishments"; even worse, the end of an "episode" is sometimes not even clearly determinable, for example, when learning to walk or writing a book about reinforcement learning. How can "intelligent" system behavior be automatically generated or optimized in this more general sense?

1.2 Machine Learning

First of all, we have to adjust our concept of "intelligence" at this point to the effect that we want to understand "intelligence" essentially as "intelligent behavior" and accordingly "learning" as an optimization of this behavior. This is consistent with the intelligent agent paradigm of AI, where any "goal-directed" behavior is said to be somewhat "intelligent". Machine learning here is understood as the generation

---

[1] "Spektrum der Wissenschaft" (issue 11/2018)

and improvement of purposeful system behaviors through iterative maximization of a "goal function". We will see that this determination makes sense especially for the field of "artificial intelligence". Through the learning process, the artificial learning procedures try to improve the outputs that an artificial system assigns to certain system inputs. For this purpose, internal representations are formed by the learning procedures in different ways, which are supposed to control the system behavior increasingly well with regard to the tasks to be performed.

In order to assess the possibilities and limitations of the various learning methods, it is useful to first classify them according to the type of feedback they receive from the environment. In general, three types of machine learning can be distinguished. They differ essentially in the way in which the "criticism" is presented, through which the behavior of the artificial system is supposed to improve. In the first variant, a "knowing" teacher corrects the system outputs by presenting the correct output, in the second, an evaluation of outputs takes place only in the form of "reward" and "punishment," and in the third, the system autonomously finds classifications and structures in the input data (Fig. 1.1).

Supervised Learning

The data stream (x_1, y_1), (x_2, y_2), , ..., (x_n, y_n) consists of pairs of inputs x with associated setpoints y for the output. In the learning process, the "training," the system produces a preliminary error-prone output. By adjusting the inner representation using the target value, the future output is shifted towards the target and the error for further outputs is reduced. Usually, the system is trained with a subset of the available data and tested with the remaining.

Important methods are the artificial neural networks with delta rule or backpropagation up to "deep" networks with numerous layers, convolutional layers, and other optimizations, which have recently become particularly prominent due to the developments in hardware. But also other methods like k-nearest neighbor classification (k-NN), decision trees, support vector machines, or Bayes classification are to be mentioned here.

Unsupervised Learning

In unsupervised learning, only the input data stream $(x_1, x_2, ..., x_n)$ is available. The task here is to find suitable representations that allow, for example, the recognition of characteristics in data sets, the recognition of exceptions, or the generation of predictions. A prior classification of the data into different classes is not necessary. However, the preselection of relevant features, as well as the corresponding "generalization function," plays a major role. "Unsupervised methods" can be used, for example, to automatically produce class divisions and keyword evaluation of Big Data. Thus, this is not about assigning patterns to existing categories but about finding clusters in a data set. Important methods are clustering and competitive learning as well as statistical methods such as dimensionality reduction, e.g., by principal axis transformation (PCA).

Reinforcement Learning

This learning method refers to situated agents that receive a reward ("Reward") $r_1, r_2, \ldots, r_n$ upon certain actions $a_1, a_2, \ldots, a_n$. Thus, here we have inputs, outputs, and an external evaluation of the output. Future actions should be improved to maximize the reward. The goal is to automatically develop the most optimal control ("policy") possible. Examples are temporal difference learning methods like Q-learning or the SARSA algorithm but also policy gradient methods like "actor-critic" methods or model-based methods.

In "supervised learning," the internal representation is adapted according to the target value, whereas in "unsupervised learning," this "teacher" is not available. Here, the representations must be formed solely from the available inputs. The goal is to find regularities in the input data, e.g., for subsequent processing, whereas in "unsupervised" learning, the algorithm has to find out the correlations—internal evaluation functions or similarity measures serve as aids—in supervised learning the correct answer, e.g., the "correct" category, is provided externally from the environment, i.e., by a teacher. Thus, in "supervised" learning methods, correct solutions are known during the training phase, e.g., distinguishing images with melanoma or harmless pigmentation. "Unsupervised" learning methods, on the other hand, can be used, for example, to optimize the spatial design of a supermarket or its pricing policy, e.g., by recognizing which goods are often bought together.

One might think that in "unsupervised learning," in contrast to "supervised learning," the system can form internal representations free of external influence. In fact, however, it must be decided beforehand in one way or another on the basis of which features and in which functional way the "clustering" is to take place, because without such a determination of relevance, the system would have no possibilities whatsoever to form meaningful divisions.

For example, apples and pears belong in the same "pot" when it comes to distinguishing fruit and vegetables but would have to be distinguished in the classification task "fruit varieties," because then apples and pears belong in different "pots"—a task-independent, quasi fallen-from-heaven ontology does not exist. In attempts to create some kind of universally applicable AI, this error in thinking has already led to some dead ends. This confirms the constructivist insight that our inner symbols and representations are not "sky hooks" fixed in the air (Dennett 2018) to which we can pull ourselves up but that our symbols and concepts are ultimately generated by the manifold problems we face, e.g., as social cultural beings.

This has implications for our notion of "objective truth." At this point, I would like to get a little philosophical for a moment. From the perspective of an intelligent agent capable of learning, cognitive representations exist because they produce useful distinctions, i.e., ultimately because they produce meaningful differences in behavior. They are not "objective" and true in themselves. Generalized, this would also apply, for example, to the laws of nature or propositions, for example, "Climate change is real and man-made." The proposition becomes true when one determines, on subjective motives, that it is reasonable to take it to be true. Sometimes cognitive representations are useful to a large group of "agents" because they are comprehensible to all, are independent of individual and subjective observations, and therefore can be agreed upon, as a generally useful fact. Facts are thus true because they are comprehensible, not, conversely, presuppositionlessly "true" and therefore generally comprehensible. This reversal can have consequences, for it can inhibit innovation and creativity if expediency is unilaterally defined by "objective facts." The phrase sounds somewhat like a simple criticism of crude dogmatism, but that would only be the case if the "agent" could not take note of empirical findings. Rather, the constructivist argument is that observations not only confirm or disconfirm hypotheses and models but that hypotheses and models equally "construct" the observation and influence the choice of the object of attention. In an active "open" inquiry, hypotheses/models and observations mutually construct and influence each other.

Some AI researchers sometimes pursue the idea of "omniscient," universally applicable "Artificial General Intelligences" (AGI). Does the assumption that realities are subjectively constructed mean that such universally applicable artificial intelligence cannot exist in principle? This is certainly not the case, just as the impossibility of a "perpetual motion machine" does not prove that automobiles do not exist. However, some preconditions must be taken into account. In reinforcement learning, the generation of intelligent behavior necessarily requires time and energy for exploration and is always related to more or less specific tasks.

If the generation and adaptation of internal representations in "unsupervised learning" take place according to an internal error function, which was previously anchored from the outside, but then does without the support of an external teacher, then we also speak of "controlled learning." If at least a part of the error signal, i.e., the feedback, originates from the environment of the system, then there is consequently an evaluation of the output by a kind of external reward which gives us the setting of "reinforcement learning." Here, the evaluation of actions can also be done by simulations in a, perhaps self-constructed, model, e.g., in a board game, by results in simulated games. Here, the algorithm for training can also use earlier versions of itself as an opponent.

In David Silver et al. (2021) argue that the reinforcement learning setting is sufficient to generate "General Intelligence" and its various manifestations: "A sufficiently powerful and general reinforcement learning agent may ultimately give rise to intelligence and its associated abilities. In other words, if an agent can continually adjust its behavior so as to improve its cumulative reward, then any abilities that are repeatedly demanded by its environment must ultimately be produced in the agent's behavior. A good reinforcement learning agent could thus acquire behaviors that exhibit perception, language, social intelligence and so forth, in the course of learning to maximize reward in an environment, such as the human world, in which those abilities have ongoing value." While they point out that "the perspective of language as reward-maximisation dates back to behaviourism," at the same time, they underscore that the problem of reinforcement learning substantially differs from behaviorism "in allowing an agent to construct and use internal state." Admittedly, however, the human brain is extremely optimized in terms of task domain size and corresponding resource consumption, so it is incredibly flexible and highly efficient, and a far cry from what we manage with our computers today.

In order to process feedback, especially from large or complex state spaces, a reinforcement learning algorithm can also make use of supervised learning techniques. A particularly prominent example of a combination of reinforcement learning with deep learning is the AlphaGo-Zero system mentioned above, which initially knew nothing but the rules of Go. In the meantime, the machine plays significantly better than the best human Go players worldwide. A similar system for chess is called "Leela Chess Zero." The game of Go is significantly more complex than chess, and it was long believed that it would be impossible to teach a computer to play Go at a human level due to the astronomical number of possible moves—one reason why the current results are attracting so much attention. Especially when you consider that, unlike, say, "Deep-Blue" which defeated then world chess champion Kasparov, the system has come without any prior human knowledge and has only continued to improve through independent exploration of the game system, i.e., playing against itself. Nevertheless, for the sake of honesty, one must also admit that such a game board is an environment that is very clearly recognizable and simply structured compared to our everyday environment and is also easy to simulate and evaluate. In many everyday situations, such as in a business environment, the conditions are usually more complex. Nevertheless, in reinforcement learning, what is usually associated with "artificial intelligence," the self-acting development of action competencies by the machine independently of humans, emerges in a particularly impressive manner. Thus, the world's best human players in chess or Go can improve by studying the strategies produced automatically by a machine—a most remarkable and novel situation and a clear case of "artificial intelligence."

Reinforcement learning is currently one of the most dynamically developing research areas in machine learning. The results also indicate how "universal intelligence" might be achievable. In autonomous learning, the central point is not at all the generalization of observations, but an "active exploration" or a "systematic experimentation" and the associated purposeful processing of "successes" and "failures." However, this also means, as mentioned above, that this universality is not available for free, i.e., as a "gift from heaven," e.g., in the form of clay tablets with ten axioms. Exploring the agent program, i.e., adapting it to any environmental system, requires a corresponding consumption of time, energy, and computational power. Of course, it is also the case that once a machine has been learned, it can be used or reproduced any number of times. Today, for example, any person with Internet access or a very powerful computer can compete against AlphaGo-Zero. Restrictively, it should be noted here that we cannot treat "intelligence" as a social category in the context of this book but initially deal with it only on the basis of a notion of "individual intentionality" (Tomasello 2014). Unfortunately, sharing models and goals within multiagent systems with reinforcement learning can only be mentioned here in passing. Individual learning refrains from social forms, such as imitating a role model or sharing knowledge, e.g., through linguistic transmission. It could once again qualitatively change the learning process and theoretically move artificial agents significantly up a "ladder of free behavior." This field of research is actually still emerging as well.

At the level of individual intelligence, autonomous "exploration" plays an important role. Exploration, by the way, is one of the most important features that distinguishes reinforcement learning from other types of machine learning: the question of exploring an environmental system is by no means trivial, because here the trade-off between "exploration" and "exploitation," i.e., the exploration of new possibilities, on the one hand, and the most optimal possible behavior of the agent, on the other hand, must be taken into account. In this context, in order to discover new better action possibilities, the agent has to try actions that it has not chosen before, with their possible utility still unresolved. Different "explorative" and "exploitative" behaviors, as well as their expedient combination in an active, artificial agent system, are not treated in this explicit and systematic form, neither in "supervised" nor in "unsupervised" learning (Sutton and Barto 2018) treated.

By engaging in reinforcement learning, a generalized perspective on learning processes can also be gained, e.g., the preliminary output of a supervised learning system can be interpreted as direct exploratory behavior of an "actor" that a teacher rewards or punishes by presenting the desired output, where the "reward" or "punishment" is generated by the deviations from the "correct" output, through which the system control then adjusts accordingly.

In the forms of individual learning or individual adaptation, the basic pattern can be found everywhere in some way: (1) production of a preliminary "action" by the learner, (2) criticism by a feedback from the "milieu," and (3) purpose-oriented adaptation of the "acting" system (Fig. 1.1).

1.3   Reinforcement Learning with Java

The functionality of the algorithms is of course independent of concrete implementations or special programming languages. However, the reader should be able to practically understand the presented algorithms and results himself. Perhaps the reader wonders whether Java is a suitable language. In machine learning, the Python programming language is often used, which is also caused by some AI frameworks distributed by big players like Google's "TensorFlow" or the "OpenAI" platform, which have a good connection to Python. The Python scripting language has also been designed with a particular emphasis on code readability, which is why the language is popular in teaching, especially in the United States. As a result, Java frameworks have fallen behind Python alternatives in recent years. Stacks like DeepLearning4J (https://deeplearning4j.org) certainly have a strong community but have remained secondary compared to TensorFlow and PyTorch. Recently, however, things have picked up again. Very interesting in this context is the recently released open-source library "Deep Java Library" (DJL) (https://djl.ai/; 3/28/2021) which is supported by amazon—the central giant in online retail. Amazon's DJL was designed to be as easy to use and simple as possible. DJL allows machine learning models built with different engines (like PyTorch, Tensorflow, MXNet, and

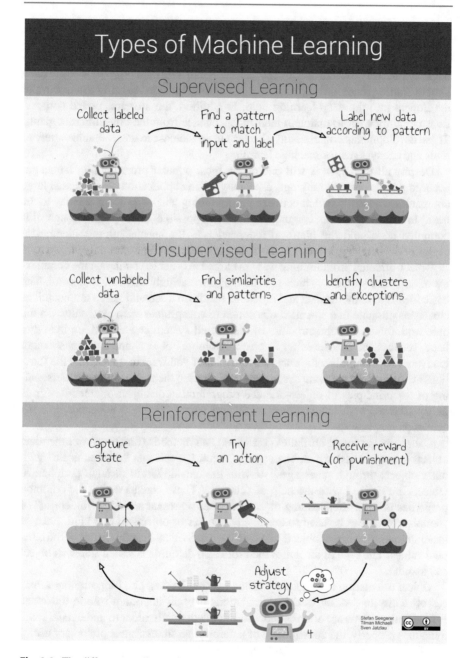

Fig. 1.1 The different paradigms of machine learning. (Image: Stefan Seegerer, Tilman Michaeli, Sven Jatzlau (License: CC-BY))

many more) to run side by side in the same JVM without infrastructure changes. Two other Java-based frameworks for deep learning that have recently been open-sourced are "Dagli" (https://engineering.linkedin.com/blog/2020/open-sourcing-dagli; 3/28/2021) from LinkedIn and "Tribuo" (https://tribuo.org/; 3/28/2021) from Oracle. Putting these tools open-source is certainly intended to bring a new wave of developers into the deep-learning field. In addition, the aforementioned corporations certainly want to maintain their independence from the other Internet giants. These developments will certainly also lead to an increase in the available variety of tools and ecosystems for machine learning.

Despite all this, Java is still one of the most popular programming languages according to the current rankings, especially among application developers at large companies who can't switch between programming languages from one day to the next. In addition to the language scope, the possible high-speed through JIT-compiled code, and the platform independence, the special, almost ideological design of the language in terms of the object-oriented paradigm certainly plays a role.

Object-oriented programming was developed in order to counteract the complexity of software systems, which is becoming increasingly difficult to control. The basic idea is to combine data and functions that form a semantic or systemic unit as closely as possible in a so-called object and to encapsulate them externally. On the one hand, this makes the structure of the overall system clearer and, on the other hand, reduces the susceptibility to errors: "foreign" objects and functions cannot inadvertently manipulate the encapsulated data. If you want to understand the functionality of modern software systems or (co-)develop them yourself, an understanding of the principles of object-oriented programming or object-oriented design is indispensable. For this reason, Java is still very popular in school education or in the corresponding courses.

Data objects contain attributes (variables) and methods (functions) or other data objects (components) which characterize them. Objects can "communicate" with other objects through "messages" or with the outside world through peripherals. Objects are defined with the help of "classes." These classes define the common properties of the corresponding objects. Classes represent a kind of "blueprint" or "template" that can be used to create any number of objects of one kind. Each of these structurally similar objects can then have individual characteristics of the attribute values. The type of attributes therefore also determines which states an object can assume.

Object orientation is excellently suited for simulations, i.e., for modeling subareas of reality or fantasy worlds. Simulations play an important role in didactics. Efficient algorithms are often "condensed simulations," in order to understand some dynamics properly, the construction of a descriptive simulation is often very useful as a first step, e.g., in neural networks. We will follow a principle of "transparency before speed" in the book. In doing so, we will transfer the discussed concepts as one-to-one as possible into corresponding base classes and implement an object-oriented design sometimes down to the individual neurons of the artificial neural networks.

Since we want to try out our adaptive programs, which we make appear in the form of autonomous agents, in simulated environments, these properties are a great asset. For the transparency of the structure, the processes in the simulation and the visualization of the states or the state changes, a language that includes object-oriented design from the ground up is very useful.

In addition, there are a large number of didactic tools for Java that enable illustrative experimentation and can also serve as an easy-to-use "playground" for testing reinforcement learning algorithms. Moreover, the language has a great similarity with C-style languages such as C++ or C#. Translating the algorithms or switching to environments with these languages is also very easy.

One such "play environment" widely used in teaching and learning is "Greenfoot." In "Greenfoot," no proprietary language is used, but "real" Java is programmed, which means that the algorithms can also be easily integrated into other applications or software projects from the Java world. Here, too, it is important to avoid unnecessary complexity without losing too much vividness. After starting the environment, a view of the current two-dimensional "Greenfoot world" appears in the central area of the screen ◉ (Fig. 1.2). The size and resolution of the world in which the objects and "actors" of the scenario are located can be changed. The software visually represents object states implicitly and without delay. As a result, manipulations to the object attributes have an immediate visual effect, and the operation of the algorithms can be directly observed.

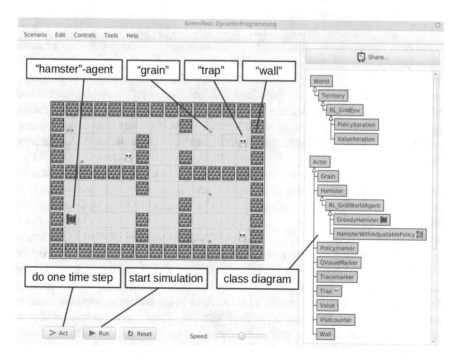

Fig. 1.2 "Hamster" gridworld in Greenfoot

The didactically optimized tool completely relieves from the work of writing graphical code for the display of elements, and one can concentrate on the development of the behavior of the "actors"—without giving up flexibility in the design of the simulated environment. It provides an immediate view of software system operations, supporting the principle of learning "by insight and experience." The system gives quick feedback and generates immediate feelings of success in short intervals. In addition, there is no need for any installation effort; the program could also be started from a USB stick. Furthermore, it should work with all operating systems on which a Java runtime environment is available, e.g., Windows, Linux, or Mac OS.

You can download it for free at https://www.greenfoot.org (April, 2020). You can also access the additional materials via the book's product page and via page (https://www.facebook.com/ReinforcementLearningJava; May, 2020) of the book you can also access materials and further information.

Curiously, such object-oriented environments are often used for teaching structured programming and when teaching basic algorithmic structures. Here, the simulated figures are commanded around in the gridworld by corresponding Java statements. However, we will undertake a much more challenging but also much more fascinating task: We will endow the figures with a life of their own during our experiments and trials and turn them into adaptive agents that independently explore their simulated environment and thus find their way around in it more and more.

Quasi-continuous state spaces can also be simulated and represented. Within Greenfoot, however, the representations are limited to two dimensions. This could also be changed, the Greenfoot GUI is based on JavaFX, if one is willing to invest the corresponding programming effort. We will use the javafx environment for additional data visualizations. It is easy to create additional windows that display more information about the running simulation.

A well suited and freely available framework for the introduction to programming artificial neural networks with Java is called "Neuroph". It can be used to add neural networks to the programs. It contains a very transparent, object-oriented, open-source Java library with a small number of base classes that directly correspond to the basic NN concepts. Which suits our principle of "transparency before speed".

There is even a GUI editor "easyNeurons" available with which networks can be designed and trained in a graphically flexible way. In reinforcement learning, ANNs can be used to estimate the evaluation of states or actions. We will deal with this later in the book.

Python fans can also benefit greatly from the explanations in the book. For applications that require a high computational effort, however, the aforementioned free high-performance libraries would also be available, which make Java attractive for professional applications as well. They take advantage of the latest frameworks for distributed computing with multi-CPUs or GPUs to support even complex training processes that occur, for example, in deep learning. These are in no way inferior to the current Python tools and also have other interesting features worth considering.

Bibliography

Dennett DC (2018) From bacteria to Bach – and back. The evolution of the mind. Suhrkamp, Berlin

Kavukcuoglu K, Minh V, Silver D (2015) Human-level control through deep reinforcement learning. In: Nature. Available online at https://web.stanford.edu/class/psych209/Readings/MnihEtAlHassibis15NatureControlDeepRL.pdf

Silver D, Huber T, Schrittwieser J (2017) A general reinforcement learning algorithm that masters chess, shogi and Go through self-play. Ed. by DeepMind. Available online at https://arxiv.org/abs/1712.01815

Silver D, Singh S, Precup D, Sutton RS (2021) Reward is enough. In: Artificial intelligence, vol 299. Wiley, Hoboken. Available online at https://www.sciencedirect.com/science/article/pii/S0004370221000862

Sutton RS, Barto A (2018) Reinforcement learning. An introduction, 2nd edn. The MIT Press (Adaptive computation and machine learning), Cambridge/London

Tomasello M (2014) A natural history of human thought, 1st edn. Suhrkamp, Berlin

Basic Concepts of Reinforcement Learning

2

> *Competence without comprehension is the modus operandi of nature, [...]. (Alpaydin 2019)*
>
> Daniel C. Dennett

Abstract

Reinforcement learning automatically generates goal-oriented agent controls that are as efficient as possible. This chapter describes what a software agent is and how it generates more or less intelligent behavior in an environment with its "policy." The structure of the basic model of reinforcement learning is described and the concept of intelligence in terms of individual utility maximization is introduced. In addition, some formal means are introduced. It is shown how interdependent states are evaluated with the help of Bellman's equation and what role the "optimal policy" plays in this process.

2.1   Agents

In the mid-1990s, the term "agent program" emerged in computer science. A definition for this was provided by Franklin and Graesser in 1996:

> An autonomous agent is a system that is both in and [at the same time] part of an environment, that perceives that environment and acts on it over time to influence what it perceives in the future with an eye toward its own goals.[1]

---

[1] In ECAI '96 Proceedings of the Workshop on Intelligent Agents III, Agent Theories, Architectures, and Languages Proceeding.

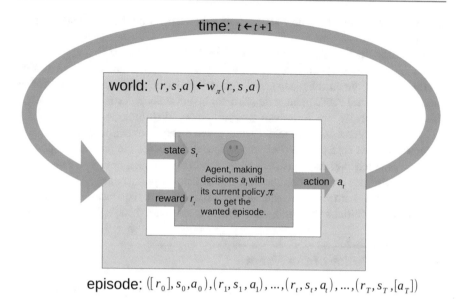

Fig. 2.1 The agent is embedded in its environment. Some processes, temporally prior to the agent's decision, influence its sensory system and provide reward signals. Under the influence of the agent's decisions, the world evolves. The agent is making decisions to get the wanted episode

This definition is very general, e.g., biological organisms can also be considered as autonomous agent systems. Agent systems are placed in an environment in which they must perform tasks. In their "life process," they must always select the best possible actions from their repertoire in the situation given to them, with regard to the tasks they have to fulfill. The world evolves under the influence of the agent's decisions and presents the agent with a new situation, which again gives rise to action decisions, cf. ◉ Fig. 2.1.

We assume that our agent has sensors that support it in making a decision based on the given "world state." The time t runs here in discrete simulation steps $t = 0$, 1, 2, ...

Let the set of all possible world states, or more precisely all possible states that the agent's "sensory surface" can assume, be denoted by S, where $s_t \in S$ let denote the state that is supplied by the agent's sensory surface at time t. Parts of the sensory surface can also refer "proprioceptively," so to speak, to the internal state of an agent, i.e., e.g., to postures, energy levels, account balances, or the like, which plays a role especially in embodied agents (e.g., robots). In our example gridworlds, a state s consists of the x,y coordinates of the corresponding cell within the gridworld and possibly other values such as the number of grains currently collected by the hamster. In board games, the state consists of the current situation of the board, i.e., the position of the pieces on it, etc.

$A(s_t) \subseteq A$ is the set of actions available to the agent for the given state and $a_t \in A$ let be the action that the agent performs at time t. After the selection of a given action, time advances by one step, and a new state s_{t+1} is reached. The reward $r_{t+1} \in R$

the agent receives and the new state s_{t+1} he finds depend on the previously chosen action, a_t action.

Regarding the mathematical modeling of the agent world, we first assume a so-called Markov system. In Markov models, the subsequent state and the reward depend only on the current state and the chosen action. However, in such a Markov model, reward and subsequent state need not be uniquely determined. To represent this mathematically, we introduce the probability distributions $p(r_{t+1}|s_t, a_t)$ and $P(s_{t+1}|s_t, a_t)$; p denotes the probability that, given a choice of action a_t in the state s_t, the reward r_{t+1} is collected in time step $t+1$, and P denotes the probability that the state s_{t+1} is reached at all, if in state s, the action a is chosen at time t. Sometimes these probabilities are also combined and described by $p(s_{t+1}, r_{t+1}|s_t, a_t)$.

If there is a defined initial and terminal state, then a sequence of actions that move from the initial state s_0 to a final state s_T is also called an "episode" or "trial."

2.2   The Policy of the Agent

From the set of all possible actions $A(s_t)$, the best possible one must be selected. This task is performed by the policy π with $a_t = \pi(s_t)$ (for deterministic policies). The policy π assigns to the states s of the set $\mathbf{S}$ an action a from the set $A(s_t) \subseteq \mathbf{A}$ $\pi : S \to A$. Thus, it defines the behavior of the agent. For the word "policy," we will sometimes also use the words "tactics" or "control." "Tactics" is more reminiscent of acting in a board game, while "control" may be more associated with situational behavior within a systemic context, such as vehicle control. In principle, however, the terms are synonymous. Later, we will introduce generalized stochastic policies that provide a choice probability distribution over the available actions.

A central problem in reinforcement learning is that often only very few actions provide a real reward, i.e., between the initial state and a state in which the agent can realize a reward, there are often very many actions or states that initially yield nothing or only cause effort, i.e., "negative rewards."

For each policy π, there is an expected cumulative reward $V^\pi(s)$ that the agent would receive if this tactic from the state s_t would be followed. Thus V^π assigns a real value to each state depending on the current policy. To calculate the value of a state, we need to sum up the rewards of the subsequent states achieved by the actions chosen by π. In the "finite horizon model" or the "episodic model," we simply sum up the rewards of the next T steps (● Fig. 2.2).

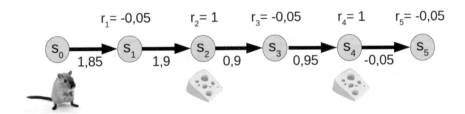

Fig. 2.2 Evaluation of actions in the finite horizon model

$$V^{\pi}\left(s_{t}\right) = E\left[\sum_{i=1}^{T} r_{t+i}\right] = E\left[r_{t+1} + r_{t+2} + r_{t+3} + \ldots\right] \qquad (2.1)$$

As a rule, however, shorter paths, i.e., target states close to the agent, are to be favored. Often, therefore, it is better to collect a potential reward in a timely manner, in keeping with the adage "A bird in the hand is worth two in the bush." This is especially true when we are in a risky environment where negative surprises can occur. For this purpose, a discount rate $0 \leq \gamma < 1$ is added. The more γ approaches 1, the more "farsighted" the agent becomes (⦿ Fig. 2.3).

$$V^{\pi}\left(s_{t}\right) = E\left[\sum_{i=1}^{\infty} \gamma^{i-1} r_{t+i}\right] = E\left[r_{t+1} + \gamma r_{t+2} + \gamma^{2} r_{t+3} + \ldots\right] \qquad (2.2)$$

(Model of the "infinite horizon"). [2]

To illustrate the effect of γ, imagine that in a labyrinth there is a "tidbit" from which an intense "smell" emanates into the corridors of the labyrinth. The farther away the source of the odor, the paler the quantitative perception of the "value" becomes. A value close to 1 would mean that an agent would give equal consideration to very distant rewards. A hypothetical value greater than 1 would lead to infinite valuations occurring, and we would have a sort of deflation problem: "The longer I wait, the better."

We now want to find a policy whose value is maximal ("optimal tactics"). With this policy, we could obtain the maximum achievable utility V^{*} for each environmental state s:

$$V^{*}\left(s_{t}\right) = \max_{\pi}\left[V^{\pi}\left(s_{t}\right)\right], \forall s_{t} \qquad (2.3)$$

$\gamma = 0{,}9$

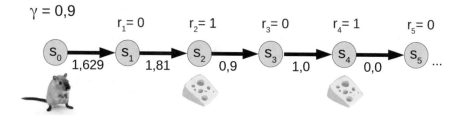

Fig. 2.3 Evaluation of actions with infinite horizon and discount $\gamma = 0.9$

[2] cf. Alpaydin (2019).

2.3   Evaluation of States and Actions (Q-Function, Bellman Equation)

The function $V(s_t)$ represents to the agent how beneficial it is to be in a certain state s_t. As we will see later, for some scenarios or learning and exploration strategies, it is useful not only to record state values but also to store what value it has to choose a certain action a in a state s . This is done by the so-called Q-function (Watkins 1989): $Q(s_t, a_t)$.

We can now define that the optimal value of an action a should be determined according to what "benefit" it brings if this action is chosen in state s and then the optimal policy π^* is followed. We call the corresponding function $Q^*(s_t, a_t)$. If we assume a purely "greedy" choice of action, then the maximum value over the various choices of action comes from $Q^*(s_t, a_t)$ of the state valuation $V^*(s_t)$ cf. Fig. ◉ 2.4.

If we now add up the expected action rewards for each state s_t and examine the terms, we find that the valuation of a state is given by s_t is the value of the best subsequent state, plus the reward that is directly realizable at the transition:

$$V^*\left(s_t\right) = \max_{a_t} Q^*\left(s_t, a_t\right)$$

$$= \max_{a_t} E\left[\sum_{i=1}^{\infty} \gamma^{i-1} r_{t+i}\right]$$

$$= \max_{a_t} E\left[r_{t+1} + \gamma \sum_{i=1}^{\infty} \gamma^{i-1} r_{t+i+1}\right]$$

$$= \max_{a_t} E\left[r_{t+1} + \gamma V^*\left(s_{t+1}\right)\right]$$

(2.4)

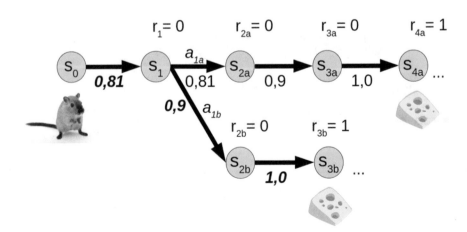

Fig. 2.4 Condition assessments for alternative action paths

This is a very useful insight, since it implies that to estimate the value of a state in the deterministic case, we only need to know the valuation of the best neighboring state.

We can generalize this to probabilistic cases without much difficulty, which leads us to the well-known Bellman equation:

$$V^*(s_t) = \max_{a_t} \left(E[r_{t+1}] + g\sum_{s_{t+1}} P(s_{t+1} | s_t | , s_t | , a_t) \bullet V^*(s_{t+1}) \right) \qquad (2.5)$$

The above provisions are based on a notion of "intelligent behavior" as "individual utility maximization." It is useful to reflect on such basic presuppositions. Our notion of "intelligence" is inadequate and open to criticism in many ways. The above model has provided us with a first basis to define "optimal behavior" and also provides us with a first handle for concrete experiments. Before we deal with actual learning algorithms, we will first look at the computation of such an optimal action strategy and illustrate the operation and principles of the corresponding algorithms. We will experiment with two scenarios for which, on the one hand, we can compute optimal strategies—indeed, this is often not even possible for most board game scenarios in a computational time available to ordinary mortals; on the other hand, they are also complex enough to investigate automatic learning later. We can then compare the strategies produced by the algorithms with the optimal controls.

What would the figure ◉ Fig. 2.4 look like for a finite and deterministic model without discounting but with movement costs of $r = -0.05$? The solution is shown in Fig. 2.5

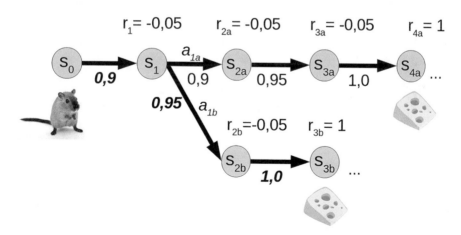

Fig. 2.5 Evaluations in a finite and deterministic model with movement cost of $r = -0.05$

How would the transition values in Fig. 2.4 change under an indeterministic transition model in which the probability is 0.8 that the desired action is performed but the probability is 0.2 that an alternative, if any, is selected? (Fig. 2.5)

It is valid with this information $p(s_{2a}|s_1, a_{1a}) = 0, 8$ and $p(s_{2b}|s_1, a_{1a}) = 0, 2$; therefore

$$Q\left(s_1,a_{1a}\right) = 0,8 \cdot g \cdot 0,9 + 0,2 \cdot g \cdot 1,0 = 0,828$$

Because moreover $p(s_{2a}|s_1, a_{1b}) = 0, 2$ and $p(s_{2b}|s_1, a_{1b}) = 0, 8$ would be

$$Q\left(s_1,a_{1b}\right) = 0,8 \cdot g \cdot 1,0 + 0,2 \cdot g \cdot 0,9 = 0,882$$

$$V^*\left(s_1\right) = \max_{a_t} Q^*\left(s_1,a_{1x}\right) = 0,882 \quad ; x \in \{a,b\}$$

The best action would still be $\pi^*(s_1) = a_{1b}$. However, the values have converged a little.

Bibliography

Alpaydin E (2019) Machine learning, 2nd expanded edition
Dennett DC (2017) From bacteria to Bach and back. The evolution of minds. Norton, New York
ECAI '96 Proceedings of the workshop on intelligent agents III, agent theories, architectures, and
    languages proceeding
Watkins C (1989) Learning from delayed rewards. King's College, Cambridge

Optimal Decision-Making in a Known Environment

3

> *The question whether objective truth can be attributed to human thinking is not a question of theory but is a practical question. Man must prove the truth — i.e. the reality and power, the this-sidedness of his thinking in practice. The dispute over the reality or non-reality of thinking that is isolated from practice is a purely scholastic question. From "Theses on Feuerbach"*
>
> Karl Marx

Abstract

This section describes how to compute an optimal action strategy for an environment with a finite number of states and action possibilities. You will learn the difference between an off-policy and an on-policy evaluation of state transitions. Value iteration and iterative tactic search techniques will be introduced and applied and tried in practice scenarios using the Java Hamster. Iterative tactic search, as a mutual improvement of evaluation and control, is introduced as a generalizable strategy for finding optimal behavior. Furthermore, the basics of computing optimal moves in a manageable board game scenario with adversaries are described.

Supplementary Information The online version contains supplementary material available at [https://doi.org/10.1007/978-3-031-09030-1_3].

If one has complete access to the states of the environment and the possible transitions, for example, in a board game, or an exact model which describes this environment completely, then basically nothing needs to be explored. We have a map, and we also know at which points we can get which "rewards." So we know under which conditions the agent gets into which subsequent state and whether we get a reward or not. What we do not know, however, is a path that leads to the greatest possible reward at the least possible cost. In our "navigation system," we have a comprehensive map, so to speak, but no route planning.

In such a state space, we can also think of board game situations or robotic controls, the subset of states where we actually get a success message ("reward") is usually very small compared to the size of the whole state space. What we lack is an assessment for the large number of "intermediate states" that may lie on our path to the goals. Thus, we need an evaluation of all states in such a way that it results in the most optimal possible action guidance for our agent.

Our "map," i.e., the state space described by our model, can have enormous dimensions. There are also cases, such as in the board game Go, that although we have an exact model of the system that can predict exactly what will happen for a given action and what reward we will receive, the state space is so large that it is not possible for us to examine all states. In this section, we first assume that it is possible for us to fully process this state space.

How we deal with unknown environments or with environmental systems that are so large that we cannot fully assess them with usually available resources will then be addressed starting in ▶ Chap. 4.

3.1 Value Iteration

3.1.1 Target-Oriented Condition Assessment ("Backward Induction")

Such a state evaluation is provided by an algorithm called value iteration (Richard E. Bellman 1957) or backward induction. For each state, the evaluation is repeatedly updated by means of the values of its neighbors given at the time. This is a problem decomposition according to the "principle of dynamic programming." In "dynamic programming," an optimization problem is solved by decomposing it, if necessary recursively, into subproblems for which optimal intermediate results can be found on a simpler level with which the main problem can be solved. The prerequisite for this is that the main problem consists of many similar subproblems and that an optimal solution results from the optimal solutions of the subproblems.

The transition from a state s with the action a to a neighboring state s' is evaluated with $E(r|s,a) + \gamma V(s')$. The E stands for the "expected value." In the deterministic case, the highest valued transition possibility is used for the calculation of the state value.

Value iteration is shown to converge to the correct values for V^* (Russell and Norvig 2010). Helpfully, what ultimately matters to us is not the exact magnitude of the valuations, but only the sequences of actions that imply the corresponding tactics. With each update step, a state score gets better or it stays the same. Often the tactic becomes optimal before the state ratings have converged to their correct values. The difference between the cumulative benefit with the preliminary tactic and the benefit of an agent that would use optimal control reduces toward zero over the course of value iteration. If the maximum change in valuation observed over the course of an iteration step is less than some threshold δ, then we say that the values $V(s)$ have converged and we stop the iteration loop.

Pseudocode "Value-Iteration"

```
1     Initialize V(s) arbitrarily for all s ∈ S (e.g. 0); V(s_terminal) = 0

2     repeat

3         Δ← 0

4         for all s from S

5             v_old ← V(s)

6             for all a from A(s)

7                 Q(s,a) ← E(r|s,a) + γ V(s')

8                 V(s) ← max Q(s,a)
                          a

9             Δ← max(Δ, |v_old − V(s)|)

10    while Δ > δ  (δ small threshold value for the detection of convergence)
```

The update rule presented in the listing applies to the deterministic case. In the case of only uncertainly predictable subsequent states or rewards, we have to consider the corresponding probability distribution of the subsequent states in line for the computation of the action evaluations, as already presented above for Markovian decision processes. In line 7 we must then use $Q(s,a) \leftarrow E(r|s,a) + \gamma \sum_{s'} p(s'|s,a)V(s')$.

Example "Hoarding"

For our experiments and exercises, we first use a simple dynamic setting with a mobile agent "Java-Hamster-Model" (Bohles) and a small game board scenario "TicTacToe." For these simple scenarios, we can easily understand the behavior of the algorithms. Moreover, the optimal state evaluations in these scenarios can be fully computed not only theoretically but also in practice. This allows us later to contrast the tactics determined by automatic learning with the optimal policy. In addition, the examples are also intended to encourage "play" and exploration. The basic ideas in reinforcement learning are often not as complex as some sophisticated practical implementations or mathematical explanations might suggest at first glance.

The scenario of an agent hoarding grains can be generalized well, so we can also easily think of any game character or vacuum-cleaning robot, etc. However, the "greed" for rewards, which plays a major role in reinforcement learning, is very nicely illustrated with the hamster, which also predestines the model for didactic use. On the website of the hamster model (https://www.java-hamster-modell.de , 11.4.2020), there is a "hamster simulator" especially created for programming didactics with the "Java hamster." However, we use an implementation in the "Greenfoot" (https://www.greenfoot.org, 20.2.2020) development environment.

After you have started the Greenfoot environment, you can load the corresponding scenario via the menu Scenario>Open. To do this, go to the folder "chapter3 optimal decision making\DynamicProgramming." When you go to Open, the corresponding scenario should start.

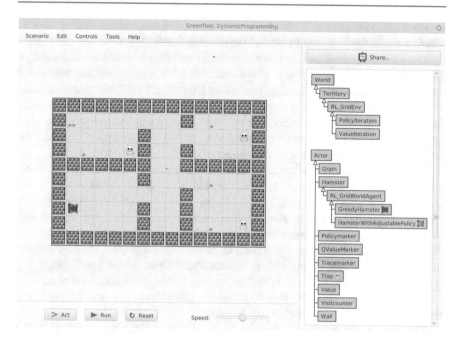

Fig. 3.1 Greenfoot dynamic programming scenario with the Java Hamster

An important role for further experiments and changes is played by the class diagram on the right side, "Worlds" are derived from the class "World," while "Figures" or other more or less movable elements in the gridworld inherit from the class "Actor." Under the context menu item "new...," you can create corresponding objects and place them on the grid. With "Open editor" you can also display the Java code of a class. If you want to see the image as in ◉ Fig. 3.1 with the red hamster after opening the scenario, you would have to start the environment manually via the context menu at the "value iteration" field in the class diagram by selecting "new value iteration."

The rewards in the "hamster scenario" are generated from the grains distributed in the environment (r = +1). Walls are impenetrable for the agent. Damage is caused on fields with traps ("Reward" r = −10).

The hamster has four degrees of freedom on the fields, but they are not always available if, for example, it is on a wall or on the edge of the gridworld. We number the actions as follows:

$$a_0 = "\textbf{\textit{Nord}}"; \ a_1 = "Ost"; \ a_2 = "Süd"; \ a_3 = "West"$$

In Greenfoot you can display the current attribute values of the objects on the field by right-clicking on the object (then on "inspect") and also call methods by selecting the corresponding functions in the context menu, e.g., take() to pick up a grain. If there are several objects on one field, "right click" will first display a selection of the available objects. Then the methods are listed according to the inheritance hierarchy. This can also be quite useful, for example, if you want to try out the effect of a code sequence. With "Act" a simulation step can be executed, while with "Run" a repetition loop can be started (speed with slider "Speed").

The scenario is designed in such a way that first the iteration is executed and then the hamster runs off accordingly. An episode ends when a grain has been reached. If the hamster is to continue running, you would have to call the "void take()" method on the hamster via the context menu. This will pick up the grain and recalculate the field weighting. After that, the hamster will start running again to the next grain. You can also place additional grains beforehand by creating additional grains with the constructor new grain() in the context menu of the corresponding class.

On the Gridworld fields, there are still objects distributed, which you probably cannot see, because they must be made visible first, "Tracemarker" and "Value" objects. Tracemarkers will only play a role in the section on tactical search. The "Value" objects allow you to view the current state scores "live." You should definitely try this out to be able to observe the development of the state evaluation.

To display the state ratings, you need to do the following: go to the Java code of the Value class (context menu of the class and then "Open Editor") and change "private boolean isOutputAn = false;" in "private boolean isOutputAn = true;"

After translating the code, blue "0.0" values should have appeared on the fields. By pressing "Run" you start the value iteration.

The state evaluations are stored per se in a two-dimensional array in the "ValueIteration" class. However, on iteration, this is mirrored into these field value display objects of the "value" class. This is done by the method "void updateDisplay()."

In the course of the iterations, the ratings propagate starting from the states with direct reward. It is interesting to note how the ratings "flow" into spaces without reward states, cf. ◉ Fig. 3.2.

The parameter γ which still appears here as an attribute of the environment system reduces the influence of distant rewards on the evaluation of a state. You will find it in the class "Value iteration" (or "policy iteration") with the default value 0.9. We will implement the parameter γ later as a property of the agent or its learning algorithm, which basically represents how important it is to collect rewards that are as close as possible. Feel free to test a smaller value, e.g., $\gamma = 0, 8$. You will be able to notice that then the hamster in the arena "apartment" gives preference to the target with the smaller reward +1 in the "next room" over the more distant target with the larger reward +2.

In the RL_GridWorldAgent class, you can set a different "transit model." This allows you to model that the agent reaches neighboring states only with a certain probability, whereby an action can no longer be assigned a unique subsequent state, but a probability distribution is formed over the possible subsequent states. To do this, you must enter the following in RL_GridWorldAgent in the line with

```
protected static double[][] transitModel = transitDeterministic;
```

select one of the other options, e.g.

```
protected static double[][] transitModel = transitRusselNorvig;
```

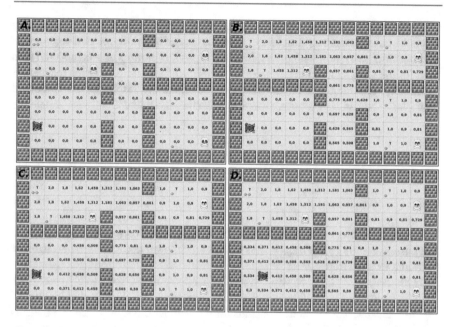

Fig. 3.2 Value iteration with deterministic transition and $\gamma = 0.9$. Left-top. A. Initial state, right-top. B. State after one iteration, left-bottom. C. Evaluations after five iterations. D. The stop state ($\delta_{min} = 0.001$)

In this model, the agent has an 80% probability of reaching the favored state but goes left 10% of the time and right the remaining 10% of the time. This model replicates the example from (Russell and Norvig 2010) after. You can also read the behavior from the matrices above or add your own transition model. If you choose an indeterministic transition model, then the scores change significantly, especially near the traps, since there is some probability of falling into the traps. If you select the proposed 10:80:10 model, then you can see that the hamster in the "apartment" again prefers the closer grain on the lower right, while with deterministic transition, on the other hand, it heads for the more distant "double grain" on the upper left.

Experiment with a transition model "transitChaotic," where the hamster reaches the front, the left or the right state with 1/3 probability each. What adjustments to the code must be made for this?

In the RL_GridWorldAgent class, a structure must be added in the form

```
public static final double[][] transitChaotic = {
                {0.34,0.33,0.0,0.33},
                {0.33,0.34,0.33,0.0},
                {0.0,0.33,0.34,0.33},
                {0.33,0.0,0.33,0.34}};
```

and be inserted in the corresponding line directly under the declarations of the transit models:

```
protected static double[][] transitModel = transitChaotic;
```

There should be worse valuations near the negative states ("traps"), especially when the hamster cannot "turn away," because instead of the discounted value of the best s', which we reach with probability 1, the product sum $\gamma \sum_{s' \in S} P(s'|s,a) \cdot V(s')$ is used for the state valuation.

Initially, the default is an environment that remotely resembles an apartment floor plan. You can also set other maps by selecting the line with

```
super(mapFlat);
```

e.g., replace with

```
super(mapWithTrap1);
```

Figure 3.3 shows the effect of an indeterministic transition model. The short path that passes directly by the traps is thus valued much lower because of the probability of falling into the trap.

In the class RL_GridEnv, you can see the available maps. You can also see in the code how the maps are defined using ASCII characters, "M" here stands for "wall," "H" for hamster, and "1" and "2" stands for the number of grains placed on the corresponding field respectively. "F" stands for trap. You can also create your own maps or mazes in this way.

In other maps should, if there is no escape route for the hamster, e.g., in a narrow corridor, the negative influence of the traps with indeterministic transition models increase significantly. In addition, it should then also take significantly longer until the iteration has converged.

Definition Often the Map "mapWithTrap1"

```
protected static final String[] mapWithTrap1=
                    {"000000",
                     "000000",
                     "0MMMM0"
                     "000000",
                     "HMFFM1"};
```

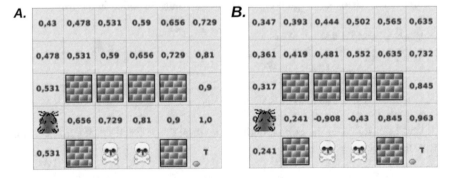

Fig. 3.3 Comparison of the evaluations with deterministic transition on the left (**a**) and probabilistic model on the right (**b**)

A vivid image, e.g., for classroom applications or introductory courses, is the metaphor of an "odor" spreading from pieces of cheese in a maze containing hungry mice. It is recommended that in the school setting, we initially limit ourselves to the deterministic case and not use transition models. In the materials for the book, which you can find on the book's product page, there is also a scenario without the probabilistic transition models. It also contains some programming tasks. In Greenfoot, tasks can be stored in the code with the string /*# ... */. These are then colored pink in the editor. This feature is suitable, for the preparation of own teaching materials.

The Java code that performs a full "sweep" of the value iteration, i.e., for each field s of the hamster territory S (which is not a wall or the terminal state), the field value $V_k(s)$ can be found in the following listing.

Value Iteration in JAVA

```
private boolean evaluateStates(){
 List objects; double maxDelta=0.0; for ( int i=0;i< worldWidth;i++ ){
    for( int j=0;j< worldHeight;j++ ){
        objects = this.getObjectsAt(i,j,Wall.class);
        if (( objects.size()==0 ) && (! isTerminal(i,j))){
            double v_alt = V[i][j];
            Actionvalue maxAW = targetOrientedEvaluation(i,j);
            V[i][j] = maxAW.v;
            double delta = V[i][j]-v_alt;
            if (maxDelta<delta) maxDelta = delta;
        }
    } } return (maxDelta<=minDelta);
}
public Actionvalue targetOrientedEvaluation(int x, int y){
  double  maxV=Double.NEGATIVE_INFINITY;   int  maxA=-1;  double
value=0;
 List <Integer> A = coursesOfAction(x,y);
 for ( Integer a : A ){
    ArrayList <Transition> successorStates =.
        GreedyHamster.successorStateDistribution(this, x,y,a);
    value = weightedValuation(successorStates);
    if ( value> maxV ) {
        maxV=value;
        maxA=a;
    } } return new Actionvalue(maxA,maxV);
}
```

The evaluation of the neighboring states is done by the function "targetOrientedEvaluation." We will later distinguish it from a "policybasedEvaluation." The weighted evaluation according to $\sum_{s',r} p(s',r|s,a)\left[r+\gamma V(s')\right]$ for the indeterministic case is done by the function "weightedValuation." It gets as argument the probability distribution of the subsequent states, which is determined by the transition model of the hamster agent.

```
public  double  weightedValuation(ArrayList  <Transition>  subse-
quentStates){
  double v = 0.0;
  for (Transition t : subsequentStates){
     v+=t.p*(getReward(t.neighborX,t.neighborY)+
     GAMMA*getV(t.neighborX,t.neighborY));
  }
  return v;
};
```

In our hamster scenario, the class "ValueIteration" contains the two-dimensional array double[][] V, which contains the current state of the state evaluations $V_k(s)$ where k stands for the number of iterations performed. With the method getV(x, y), the current values can be queried.

It is technically obvious to use two arrays, one containing the old values of $V_k(s)$ and one for the new values $V_{k+1}(s)$. This would allow the new values to be computed cleanly separately, using the old results, without having both old and new values in the same data structure at the same time, and the computation being based on old and new values in random proportions. Depending on the order in which the states are updated, sometimes new values are already used instead of the old ones. This is harmless, however, since this so-called "in-place algorithm" also leads to V^* converges (Sutton and Barto 2018, p. 75). It is usually even faster than the two-array version, because it uses the new data immediately, as soon as they are available. For the "in-place algorithm," however, the order in which the states are updated during the run has a clear influence on the speed in which the values converge.

For the understanding of the dynamics of the algorithm it is informative to observe how in the course of the iterations, starting from the states with high evaluation or directly accessible reward, the "affordances," the high evaluations spread in all directions over the entire attainable state space. For this, the states should be low or initialized to 0. The values melt through the discount factor with increasing distance from the source of the value. At points where two different valuations meet, the higher valuation prevails. If a valuation meets the state the agent is in, then a goal-directed episode could be determined. The algorithm determines quasi starting from the attractive target states in the state space in all directions the ways up to the state of the agent. What is also underlined by the term "backward induction."

Control-based solely on state assessments $V^*(s)$ and a simple "greedy" policy may be inadequate, however, as the following example shows. A "greedy" tactic that always directly follows the highest-valued neighbor state can lead to dangerous situations, as the example in ◉ Fig. 3.4 shows. Such an agent here would ignore the danger posed by the "traps." Although we have calculated probabilistically, the rewards to be expected in the long run are nevertheless lower when following the

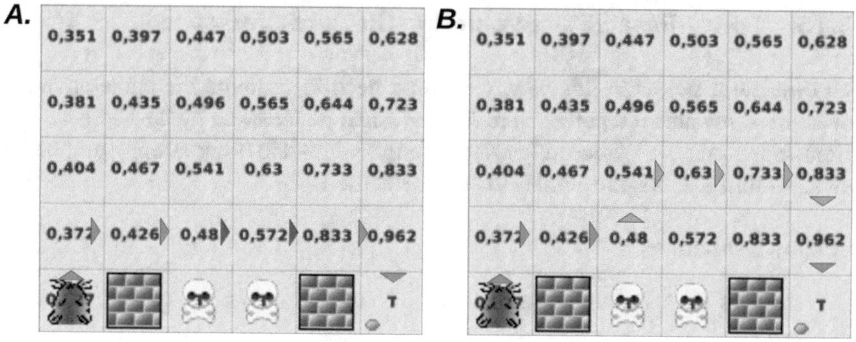

A.

0,351	0,397	0,447	0,503	0,565	0,628
0,381	0,435	0,496	0,565	0,644	0,723
0,404	0,467	0,541	0,63	0,733	0,833
0,377	0,426	0,48	0,572	0,833	0,962
					T

B.

0,351	0,397	0,447	0,503	0,565	0,628
0,381	0,435	0,496	0,565	0,644	0,723
0,404	0,467	0,541	0,63	0,733	0,833
0,377	0,426	0,48	0,572	0,833	0,962
					T

Fig. 3.4 (a) progression with "greedy" tactics based on state ratings. (b) an optimal episode

riskier path than when following the safe path (● Fig. 3.4b). This problem also appears in the literature as the "windy cliff walking" scenario (Sutton and Barto 2018).

According to $Q(s,a) \leftarrow \sum_{s',r} p(s',r|s,a)[r + \gamma V(s')]$, the value of 0.48 in state (2;3) arises from a possible transition with a_0 to "up" to (2;2).

$$Q([2,3],a_0) = g(0,1 \cdot 0,426 + 0,8 \cdot 0,541 + 0,1 \cdot 0,572) = 0,47934 \approx 0,48$$

Similarly, the value 0.572 in (3;3) arises from the possible transition to (3;2). The actual transition from (2;3) to (3;3) per a_1 would have with our transition model, however, only a value of:

$$Q([2,3],a_1) = 0,1 \cdot (-10) + g(0,1 \cdot 0,541 + 0,8 \cdot 0,572) = -0,539...$$

which is obviously not optimal. One solution would be to make the hamster's tactics smarter and teach it to take into account its own transition model. If we use for the control $\pi(s) = argmax_a \sum_{s',r} p(s',r|s,a)[r + \gamma V(s')]$, this would allow the agent to better evaluate and select its own actions. You can try this out by replacing the P_Policy_Deterministic function in the Return statement with the P_Policy_withTransitionmodel function in the GreedyHamster's control, which is set in the P_Policy method.

```
public double[] P_Policy(int x, int y){
    return P_Policy_withTransitionmodel(x,y);
}
```

This does not change much at first glance. However, one can notice that the hamster more often follows the safer path ● Fig. 3.4b.

However, it is also possible to make a more "realistic" valuation calculation. Which in such circumstances leads to a "better," i.e., less risky, state evaluation. For this, we do not have to calculate from the desired target states but calculate what would actually happen in a situation with a given hamster policy.

3.1.2   Policy-Based State Valuation (Reward Prediction)

To evaluate a state s in this way, we would need to compute the cumulative and discounted rewards if a policy were to be π would be followed by the agent starting from state s. We also proceed here according to dynamic programming and obtain an algorithm that is quite similar to "value iteration":

"Policy-Evaluation"

```
1 Initialize Vπ(s) ∈ ℝ and π(s) ∈ A(s) arbitrarily for all s ∈ S
2 Repeat:
3      Δ← 0
4      for all s ∈ S:
5           v_old ← Vπ(s)
6           Vπ(s) ← Σ_s',r p(s',r|s,π(s))[r + γVπ(s')]
7           Δ← max(Δ,|v_old − Vπ(s)|)
8 While Δ> δ (δ small threshold value for the detection of convergence)
```

With each pass k through the state space, this method causes the values to approach V_k approach the values V^π. Do we repeat the procedure for each state s from S until the changes are sufficiently small, then we obtain an approximation for V^π. This is called "policy evaluation" in the literature. In reinforcement learning, the approach is also called "prediction problem"[1]. While there is still no difference between prediction and actual observation in our fully known environmental system, we know exactly what we will get, V^π, however, would correspond in our case to the reward we can get with our present control in the respective states. Such a more "realistic," tactics-based approach to evaluation is also called "on-policy," while the goal-oriented method is called "off-policy." We will encounter this later in the chapter on learning algorithms, which face the additional challenge of having to explore the state space.

In the hamster program, you can easily create the PolicyEvaluation by replacing the call to the function "targetOrientedEvaluation" in the iteration loop in the class "ValueIteration" with the function "policybasedEvaluation," which does the calculation with

$$V^\pi(s) \leftarrow \sum_{s',r} p(s',r|s,\pi(s))\left[r + \gamma V^\pi(s')\right]$$

done.

```
private boolean evaluateStates(){
 List objects; double maxDelta=0.0; for ( int i=0;i< worldWidth;i++ ){
     for( int j=0;j< worldHeight;j++ ){
```

---
[1] cf. Sutton and Barto (2018), Chapter 4.1.

```
            objects = this.getObjectsAt(i,j,Wall.class);
            if (( objects.size()==0 ) && (! isTerminal(i,j))){
                double v_alt = V[i][j];
                Actionvalue maxAW = policybasedEvaluation(i,j);
                V[i][j] = maxAW.v;
                double delta = V[i][j]-v_alt;
                if (maxDelta<delta) maxDelta = delta;
            }
        } }
    return (maxDelta<=minDelta);
}
public Actionvalue policybasedEvaluation(int x, int y){
    double value=0; int maxA = -1;
    double[] P = hamster.P_Policy(x,y);
    for (int a=0;a< P.length;a++){
        if (P[a]>0){
            ArrayList <Transition> successorStates =
            RL_GridWorldAgent.successorStateDistribution(this,x,y,a);
            value += P[a]*weightedValuation(successorStates);
            maxA=a;
        } }
    return new Actionvalue(maxA,value);
}
```

The class "Actionvalue" is just a kind of data record, which consists of two elements, the valuation v and the associated action a. It is only about the fact that the action, which is involved in the calculation of the maximum, is supplied by the function and is not lost.

For the following explanations, the "model-free" simple greedy policy was again set on the hamster. In a deterministic transition model with a "greedy" policy, the calculated values between "value-iteration" and "policy-evaluation" match. However, in an indeterministic transition, there are significant differences. The "on-policy" evaluations are significantly more cautious, cf. ◉ Fig. 3.5.

Why is the rating of state (3,3) so strongly negative, even though there is also a good and safe action with the transition to (3,2)? In a sense, the policy evaluation calculates in that the hamster in (3,3) with model-free greedy policy cannot resist the high valuation of 0.833 in (4,3), creating a 10% probability of falling into the lethal trap when the agent is in (3,3). At this point, the hamster's "misbehavior" is "priced in" but not remedied, freely following the inflexible "realist's" motto: "I'm just like that."

Again, a better solution would be to use a more "realistic" deliberative policy that uses the transition and environment model to calculate all the $Q(s, a_i)$ values and makes them the basis for action selection. Here the same state evaluations arise, as in the "value-iteration," the goal-oriented computation of the maximum takes place

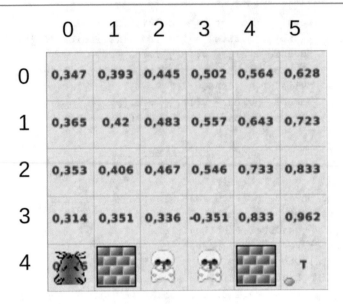

Fig. 3.5 State evaluations with the policy evaluation

now however in the policy of the hamster. The resulting behavior is the same as in the combination of "value-iteration" and "deliberative policy."

It has been shown that at the level of state evaluations we can $V(s)$ and greedy policy sometimes cannot generate optimal behavior because the value of a transition depends on several neighboring states. In the indeterministic case, to remedy this, we would have to let the hamster do some of its own reasoning about its options, or else create "finer" structures in the states and, instead of simple state evaluations $V(s)$ for each possible action option $a_i \epsilon A(s)$ in s the action values $Q(s, a_i)$ for each possible action option.

In what follows, we will change the perspective on learning algorithms somewhat.

3.2   Iterative Policy Search

In value iteration, the estimates of environmental states are updated with each sweep, which in the end affects the behavior of an agent with a more or less "greedy" control. In the tactics search approach, the other way around is to start with an arbitrary control π and improves it until there are no more improvements in the behavior.

From the point of view of an experimenting agent, we can also consider the control π as our preliminary state of practice, on the basis of which we first act spontaneously. Afterward, we reflect on consequences and possibilities for improvement. We are dealing here with an inverted perspective on the automatic development of competencies. We do not start with the evaluation of environmental states in order to induce improved behavior, but the other way around: The consequences of spontaneous action decisions are observed in order to derive improvements to the agent's control. For this, we need a much more flexible policy that allows direct assignment of actions to states. At this point, we will first solve this again in tabular form.

3.2.1 Direct Policy Improvement

The improvement of a given policy π works in such a way that by means of the evaluations $V^{\pi}(s)$ and the goal-oriented calculation $E(r|s,a) + \gamma \sum_{s' \in S} P(s'|s,a) V^{\pi}(s')$
for all possible actions in the state $a \in A$it is checked whether there is an alternative action a that would lead to a better state than the one "proposed" by our provisional control.

If this is the case, i.e., we find a better action option, the tactic is updated with the better of the action options found. It is obvious that we can be sure that the tactic will improve with each of these updates.

"Policy-Improvement"

```
1       for each s ∈ S

2              π′ ← π

3              Calculate state value "policy-based vπ = Σs′,r p(s′,r|s,π(s))[r + γVπ(s′)]

4              Check, using the values Vπ(s) and target-oriented evaluation
               for all actions a ∈ As,  va = E[r|s,a] + γ Σs′∈S P(s′|s,a)Vπ(s′)  ("value-
               based"), whether there is an alternative action a with va > vπ
               and adjust the policy accordingly:

5              π′(s) ← argmaxa(va)
```

For this we need a customizable policy where we can set which action should be chosen in a given state. This is done in the HamsterWithFlexiblePolicy class by a simple integer array:

```
protected int[ ][ ] pi;
```

The policy of the hamster is thus reduced to a query of this array. The stored action is assigned the probability 100%.

```
public double[] P_Policy(int x, int y){
    double[] P = new double[env.neighborStates.length];
    Arrays.fill(P,0.0);
    int a=getA_Pi(x,y);
    P[a]=1.0;
    return P;  }
```

A Java implementation of the "policy-improvement" algorithm is given below. For each field that is not a wall, a goal-oriented evaluation, i.e., a "greedy" check of the neighboring states, is performed. Now, however, we do not need the best evaluation, but the best action, because we are updating the policy, not the state evaluation. If the action found differs from the action currently stored in the policy for that state, the policy is adjusted accordingly. If a change occurs, the function returns false, which should mean that a stable state of Pi has not yet been reached.

Java Implementation of a "Policy-Improvement"

```java
private boolean policyImprovement(){
 boolean policystable=true;
 for ( int i=0;i< worldWidth;i++ ){
    for( int j=0;j< worldHeight;j++ ){
        List objects = this.getObjectsAt(i,j,Wall.class);
        if (( objects.size()==0 ) && (! isTerminal(i,j))){
            Actionvalue maxAW = targetOrientedEvaluation(i,j);
            if (maxAW.a!=hamster.getPi(i,j)){
                hamster.setPi(i,j,maxAW.a);
                policystable=false;
            }
        }
    }
 }
 return policystable;
}
```

This direct tactics optimization becomes particularly interesting when combined with tactics-based state evaluation. This is because these two algorithms can improve each other in a kind of "dialogue."

3.2.2 Mutual Improvement of Policy and Value Function

Both algorithms, "policy-evaluation" and "policy-improvement," are combined in the so-called "policy-iteration" (Sutton and Barto 2018). Here, again, an arbitrarily chosen policy is used to π started. This is improved by means of "policy-improvement" and the evaluation function V^π achieved so far. With the new, improved control, we then compute a correspondingly improved state score, which is in turn used to revise the policy and obtain an improved policy π'. Since the tactic is guaranteed to improve as it is adjusted, we are guaranteed to have a sequence of monotonically improving policies π_k, and corresponding evaluations V^π_k are obtained. One now repeats this until there are no more improvements.

$$\pi_0 \to V_0^\pi \to \pi_1 \to V_1^\pi \to \pi_2 \to \ldots \to \pi^*$$

Although the computation time required for policy iteration is generally higher than for value iteration, fewer runs are needed until the optimum is reached (Alpaydin 2019, Chap. 18.4.2; Russell and Norvig 2010).

"Policy-Iteration"

```
1        Initialize V(s) ∈ R and π(s) ∈ A arbitrarily for all s ∈ S

2        Repeat:

  1. evaluate states with current policy:

3            Repeat:

4                Δ← 0

5                Loop for each s ∈ S:

6                    v ← Vπ(s)

7                    Vπ(s) ← Σs',r P(s',r|s,π(s))[r + γVπ(s')]   "policy based")

8                    Δ← max(Δ, |v − V(s)|)

9                While Δ> θ  (θ small threshold value for the detection of
                 convergence)

  2. policy improvement with the updated state evaluation:

10           for each s ∈ S:

11               π' ← π

12               Check, using the values Vπ(s) and target-orientated
                 evaluation for all actions a ∈ As,
                 va = E[r|s,a] + γ Σs'∈S P(s'|s,a)Vπ(s')  ("value-based"), whether
                 there is an alternative action a with va > vπ and adjust
                 the policy accordingly:

13               π'(s) ← argmaxa(va)

14       Until no changes to policy take place at improvement step (2.) (π ≠ π')
```

To try out the algorithm in Greenfoot, you have to create the environment "PolicyIteration" (context menu of the corresponding class and then new PolicyIteration()). The hamster agent should have a green outfit after that. In the class you will find the code for a run of the policy iteration in the method "Act," which is called once when the corresponding button is pressed or repeatedly by pressing the "Run" button. You can also find the code again in the function iterate(int n), this optionally executes the policy iteration completely or with n runs.

```
public void iterate(int n){
  boolean policystabil=true; int k=0; do{.
    if ((n!=PolicyIteration.UNTIL_STABLE) && (k>=n)) break;
    boolean minDeltaReached = false; int c=0;
    while (! minDeltaReached) {
        minDeltaReached=evaluateStates();
        c++;
    }
```

```
policystable=policyImprovement();
updateDisplay();
k++; }while(! policystable);
}
```

If you press "Act" once, then the "Tracemarkers" are activated in the states, which show the current policy. With each further sweep, the ratings and accordingly the assignment of the actions are adjusted. The algorithm converges rapidly to the optimal policy within a few sweeps ⊙ Fig. 3.6 It may also be interesting to see how the algorithm provides an optimal solution for the situation at the traps, e.g., at field (3,3).

Certainly, policy iteration is not the most parsimonious solution for our scenario. For example, if we replace the "policy-evaluation" with the "value-iteration" presented in ▶ Sect. 3.1.1 presented "value-iteration," we obtain the optimally matched policy ⊙ Fig. 3.7 without much back and forth after one pass.

However, policy iteration has a special theoretical significance independent of this. Almost all reinforcement learning methods can be viewed in terms of "policy iteration" because they all involve controls $\pi(s)$ and evaluation functions $V(s)$ have, where the policy is adjusted, "greedily made," with respect to the evaluation

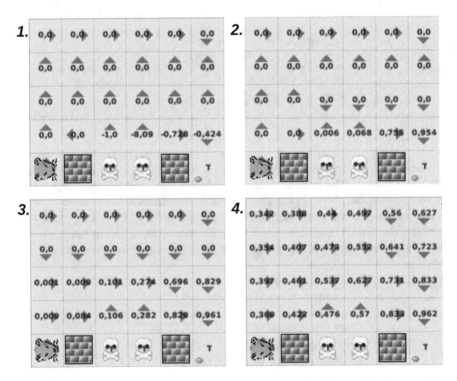

Fig. 3.6 Process of the policy iteration. Line by line from top left to bottom right the state after one, two, three, and four runs

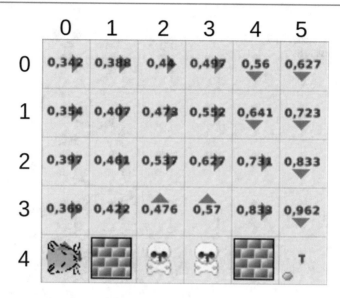

Fig. 3.7 Final result: optimal policy and evaluation

function, and the state evaluations in turn depend on the current control. In (Sutton and Barto 2018) we therefore propose the notion of a "generalized policy iteration," "generalized policy iteration" (GPI). Use the term to describe, in general terms, the idea of allowing the processes of "policy evaluation" and "policy improvement" to work together, regardless of the respective modes of realization and other details of the two processes.

Another interesting property is that it can be used to determine when the valuation function and control are optimal. The policy stabilizes only when there are no more alternative actions to choose from with respect to the given valuations—there are no more "greedy" tactics." The evaluation function stabilizes only when there are no more "shorter paths" to the rewards. The "policy improvement" moves in the dimension of action possibilities, the "policy evaluation" optimizes the paths to the goals. Selecting an alternative action affects the given evaluation of the states; on the other hand, correcting the state evaluations leads to finding new tactics improvements. The processes do not consolidate until a tactic is found that is consistent with its own evaluation that emerged from it. When both the evaluation process and the improvement process have stabilized, then the evaluation function and policy must be optimal. Incidentally, then, because of the combination of "absolutely greedy tactics" and "correct evaluation" in all states, everywhere applies $V^p(s) = \sum_{s',r} P(s',r|s,\ p(s))\left[r + gV^p(s')\right]$, the Bellman optimality criterion ► Chap. 2 is also satisfied (Sutton and Barto 2018) (Fig. 3.8).

Philosophically, we have "moved up" a bit in terms of the complexity of our conception of the emergence of intelligent behavior. Instead of considering this as a goal-oriented search for the shortest possible way, we now see this emerging from

Fig. 3.8 Mutual
optimization of tactics and
condition assessment

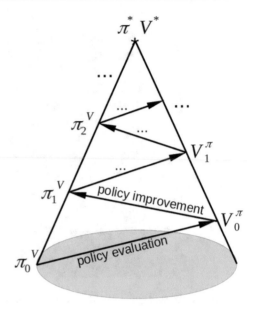

an interplay of "intuition and imagination"[2]. While "imagination" connects the states of the world in terms of their evaluations, "intuition" connects the environmental states via actions. Advances like GPI also show, in my view, how the assumption of some critics that we must remain fixed to a particular, mechanical "computerized" notion of mental activity is wrong. We are not so fixed by the construction of the computer, as some critics think, but it is the other way around that we define the structures of computational technology, even in the hardware. Even the conventional von Neumann architecture of the computer is already an extremely flexible construction material, think, for example, of weather models or the simulation of artificial nerve cells. The statement that a "mechanistic computing machine" cannot represent dialectical processes[3], for example, has as much value as the statement that a ship made of iron can never float because it is too heavy.

---

[2] This can be understood as a superficial and risky reference to Schopenhauer's philosophy. Instead of "will," "intuition" was used to suggest that this function can change. More likely, this pursuit of cumulative reward corresponds to Schopenhauer's "will." Such a reference is risky also because there is the danger of mystifying a comparatively simple, mechanical process; on the other hand, something like this can also stimulate the imagination.

[3] Processes that contradict each other, but nevertheless form a unity and may possibly result in a new level, can be canceled.

3.3   Optimal Policy in a Board Game Scenario

The findings presented can also be applied to board game scenarios. Often such games take place on a kind of grid, like chess, checkers, go or TicTacToe, etc. However, here it is now the case that the entire "gridworld" represents a state and the subsequent state set is defined by the game rules, which usually specify which moves are possible in a given situation. In addition, there is the definition of target states, which contain a more or less large profit for the players.

The states are now "temporally" adjacent and have to be unfolded with the help of a kind of simulation model. In addition, there is the hitherto undiscussed peculiarity that the environment has its own complex dynamics, which depends which subsequent state we observe after our own action. In board games we usually have to deal with an opponent who works against us, if we disregard puzzle scenarios, e.g., the game the "towers of Hanoi," various puzzles or the like. Such a board game environment suits us in that we can assume, when predicting behavior, that the opponent resembles ourselves and is simply trying to maximize his reward. Beyond that, no other independent processes independent of our actions play a role in such an "environment."

We receive a reward when we reach a winning state on the playing field. In a "zero-sum game" with two players; the opponent receives the same "reward" as we do but with a reversed sign. If the opponent wins, then we collect a corresponding "negative reward." What are the optimal tactics for such a game?

Not only do we have to find the fastest way to a winning condition, but we also have to take into account that the opponent tries to maximize his advantage after each move, i.e., to cause a worst-case for us. How do we evaluate an action possible for us under these circumstances?

Let's take a closer look at our behavior in such a game. If we can reach a winning state with the next move, then the decision is easy: we choose the action and collect the reward. What do we do if we find only action possibilities without reward? In this case we have to assume that the opponent tries to minimize our profit with his move, i.e., in this case we have to assume for every possible move that the opponent chooses an action which is best for him, i.e., which brings the smallest possible benefit for us. If the opponent cannot achieve a winning condition either, he repeats the same principle and checks what move we would choose in each case.

In ◉ Fig. 3.9, a subset of the tic-tac-toe state space is shown. There are also some ratings noted. If we can determine a winning state, then the player for whom is rated gets +100; otherwise, we take the largest of the ratings from the state cloud immediately following. However, this value must be negated when crossing the colored circles, because in a scenario with opponents, the view of the rating of a state is reversed when the player changes.

Because of this constant repetition of a similar evaluation method, we can unfold and evaluate the complete state space using a recursive algorithm. Algorithms that solve this task recursively would be, for example, the "NegaMax" or the "MiniMax." We will take a closer look at the "NegaMax" algorithm in the following. In

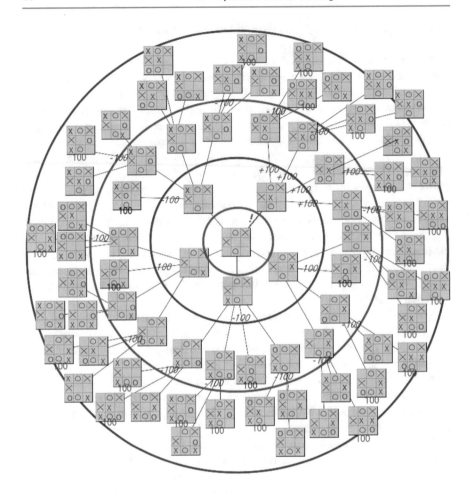

Fig. 3.9 Section of the state space in TicTacToe with an optimal move

principle, it realizes a deep search in the "game tree" and propagates found rewards
in the described way back to the top.

"NegaMax" (for TicTacToe)

```
1       function NegaMax_evaluation(s, player)  ; s ∈ S and player ∈ {'X','O'}
2       if s a terminal state
3               return Reward(s)
4       else
5               v ← −∞
6               opponent ← {'X','O'} \ player
5       for each a ∈ A(s)
6               Execute action a in s and observe s' (simulation).
7               v ← max(v, −NegaMax_evaluation(s', opponent))
8       Return v
```

If we transfer the tic-tac-toe game and the above recursive algorithm into our introduced terms, we can say that we perform a kind of goal-oriented state evaluation within a deterministic environment in which the expected reward is not discounted, i.e., our discount factor γ has the value $\gamma=1$. The algorithm calculates the optimal evaluation $V^*(s)$, because in each case we check all subsequent states, calculate the Q-values there, and choose the maximum. There are always only two cases, either we take the direct reward from terminal states or we again check all possible subsequent states by self-call. We continue this until we have reached a leaf node in each case.

Since TicTacToe is a rather concise game with only 5,478 game situations and possible 255.168 game courses, we can easily compute an optimally playing agent with our algorithm. A Java implementation of the algorithm in Greenfoot can be found in the accompanying material in the directory "Chapter 03 Optimal Decisions/ TicTacToe with NegaMax." By pressing the "Run" button, you can also compete against the algorithm yourself, while also observing NegaMax's evaluations.

The Recursive State Evaluation of NegaMax-Algorithm in Java

```
public double evaluateAction( int action, char player ){
  state[action]=player;
  double reward = getReward(state, player);
  if (reward!=0) {
     state[action]='-';
     return reward;  }
  player = (player=='o') ? 'x':'o'; // change player ArrayList
<Integer> A = coursesOfAction(state); if (A.size()==0){
     state[action]='-'
       return 0;//If the field is full,the game is over and the
return is 0. }
  double maxNegative = Double.POSITIVE_INFINITY;
  for (int i=0;i< A.size(); i++){
     double value = -evaluateAction( A.get(i), player );
     if (value < maxNegative) {
        maxNegative = value;
        }    } state[action]='-';  // undo   the   trial    return
maxNegative;
}<-!!! The Java-Code formatting, indentations etc., is very poor !!!
```

The array char[] state, which consists of nine characters, is used to store the field state currently examined by the agent. The agent simulates a move a by setting state[a]= 'x' or ='o'. To save memory, the field is defined globally. Therefore, the agent works with only one data structure, and it moves are executed "test-wise" and must be undone afterward. This is done by writing a minus sign to the corresponding position again using state[a]='-'. The minus sign stands for an empty field.

3.4　Summary

In "real-world" scenarios, we usually do not have a complete environmental model available. Furthermore, in most scenarios, it is not practical to calculate the ratings for the entire state space or to store corresponding actions because of the large number of states.

So far, we have not actually dealt with "learning" either. What we have discussed so far has actually been about producing optimal decisions based on a known model of the environment. We have discussed approaches to environment evaluation. Here we learned about algorithms of "dynamic programming." Two variants were discussed, one that evaluates the world states starting from the target states "off-policy" and another one that evaluates "on-policy," i.e., starting from the given control of the agent, which consequences a behavior would have. Furthermore, we learned about a "policy search" approach that directly converts sensory input into actions and checks whether this mapping can be improved. Finally, we combined approaches of "state evaluation" and "policy search" with "policy iteration."

The algorithms were thus based on a complete environmental model whose state space we tried to tame with the help of dynamic or recursive programming. As a rule, however, we cannot calculate through every possible state. Even simple examples give rise to an enormous number of possible states. However, if we do not have a fully evaluated model, then we have to explore the environmental system which should mean for us first of all to determine the state or action evaluations.

In the learning algorithms presented in the following, we will repeatedly encounter the principles presented in the previous sections. However, we must additionally cope with the fact that, on the one hand, we have only limited knowledge about the environment, which still has to be explored ► Chap. 4, and that, on the other hand, we have to make do with limited resources, such as computing time or memory ► Chap. 5. By overcoming these challenges, we will obtain algorithms that are significantly more powerful because they can flexibly deal with a significantly larger number of "real" environmental systems; moreover, we will gain new insights into topics such as "learning," "planning," or "decision-making."

Bibliography

Alpaydin, E (2019) Machine learning, 2nd expanded edn

Bohles D. Java-hamster-model. Available online at www.java-hamster-modell.de

Russell S, Norvig P (2010) Artificial intelligence. A modern approach, 3rd edn. Pearson Education, Inc., New Jersey

Sutton RS, Barto A (2018) Reinforcement learning. An introduction, 2nd edn. The MIT Press (Adaptive computation and machine learning), Cambridge/London

Decision-Making and Learning in an Unknown Environment

4

> *Without order nothing can exist-without chaos nothing can evolve. Nowadays people know the price of everything and the value of nothing.*
>
> Oscar Wilde

Abstract

This chapter describes how the agent can explore an unknown environmental system in which it has been placed. In doing so, he discovers states with rewards and has to optimize the paths to these goals, on the one hand, but also explore new goals, on the other hand. In doing so, he must consider a trade-off between exploitation and exploration. On the one hand, he has to collect the possible reward of already discovered goals; on the other, hand he has to manage the exploration of better paths or the discovery of new goals. There are different approaches to this; some aim at processing experiences made in such a way that the agent behaves better under the same conditions in the future "model-free methods"; and others that aim at optimizing models that can predict what would happen if certain actions are chosen.

Supplementary Information The online version contains supplementary material available at [https://doi.org/10.1007/978-3-031-09030-1_4].

The task of exploratory learning algorithms is more complex than that of algorithms that "only" have to calculate an optimal decision. This is because they additionally have to find out at which places in the state space we get rewards in the first place. We will need to send our agent on exploratory trips where it makes "discoveries." These discoveries will eventually need to affect the ratings, or behavior, in the area that has already been explored up to that point. For now, we will again solve this in a tabular fashion, ignoring the potentially huge consumption of memory. However, in the context of the simple scenarios presented above, this is easily done with today's computers. Before we do this, however, we need to look at the trade-off between "exploration" and "exploitation."

4.1   Exploration vs. Exploitation

When exploring an environment, chance must play some role in discovering new paths. A new observation usually results in the need to adjust the valuation of a state or state-action pair. It may be beneficial to adjust the existing score "cautiously" to reduce the impact of statistical outliers. The magnitude of this adjustment can be regulated using a learning rate η (often α is also used for that value). When exploring, agent control must now and then deviate from the best path determined so far. This contradicts the requirement to perform the best possible exploitation of the environment. How does one deal with this contradiction?

Methods for Probabilistic Action Selection
ε-Greedy

An obvious idea is to alternate exploration and exploitation behaviors. In the exploration phases, we perform training in which we randomly select actions and observe the results. In the exploitation phase, we act as optimally as possible according to the given estimate. It has proven useful not to completely abandon the observations collected up to that point and to run a mixed strategy. A common strategy is the so-called ε-greedy control. According to a probability specification ε, a choice is made between exploration and exploitation. A value of, e.g., $\varepsilon = 0.3$, leads to the agent being "greedy" with a probability of $1 - \varepsilon$, i.e., 70%, is "greedy," i.e., selects the "optimal" action a_{max} and 30% explores, i.e., acts randomly to make new observations. A simple Java implementation might look like this:

ε-Greedy Action Selection

```
if (random.nextDouble()<epsilon) {
    ArrayList <Integer> A =.
            (ArrayList) environment.actionPossibilitiesFor(s);
    return A.get(random.nextInt(A.size()));
} else {
    return getActionMaxQ(s);
}
```

In probabilistic algorithms, however, our policy does not return a single action, but a probability distribution $P(a|s)$ over the possible actions of the agent in a state s; such policies are also called "stochastic policies."

If $s \in S$ is an arbitrary state and $A(s) \subseteq A$ is the set of actions available to the agent for the given state, then for ε-greedy choices of different actions, compute $a \in A(s)$; the probability distribution is given by:

$$P(a|s) = \begin{cases} \dfrac{\varepsilon}{|A(s)|} + 1 - \varepsilon & \text{if } a = a_{max} \\ \dfrac{\varepsilon}{|A(s)|} & \text{for all other actions } a \end{cases} \qquad (4.1)$$

where $|A(s)|$ denotes the number of possible actions in state s.

In more complex environments, it makes sense to vary the amount of exploration, e.g., to start with a high epsilon and to continuously reduce the amount of exploration in the course of the learning process.

SoftMax

An other "soft" control can also be implemented in such a way that it uses the spectrum of neighborhood evaluations not only in such a way that either the best action is picked or not but that we produce a probability distribution over the different possible actions according to the evaluation estimates produced so far. Conveniently, the probability of selecting an action a in state s is always greater than 0. The so-called SoftMax function can serve for the realization of such a task.

$$P(a|s) = \frac{e^{Q(s,a)}}{\sum_{b \in A}^{A} e^{Q(s,b)}} \qquad (4.2)$$

Here we can also use a kind of "cooling strategy," where we introduce a temperature variable T.

$$P(a|s) = \frac{e^{Q(s,a)/T}}{\sum_{b \in A}^{A} e^{Q(s,b)/T}} \qquad (4.3)$$

For small T better actions are preferred; for large T the probabilities converge, and we get increasingly exploratory behavior. The strategy of starting with a large T and then continuously reducing it is also called "annealing." With which we can produce a smooth transition from exploration to exploitation.

Sometimes it is important that none of the actions possible in a state is assigned probability 0, which would completely exclude the action possibility in question. Controls for which it is true that $\pi(a|s) > 0$ for all $s \in S$ and all $a \in A(s)$ are called "soft-policies."

In more complex environments, we often encounter the problem that we can only collect extrinsic rewards very sparsely. This leads to the need to optimize exploratory behavior and to search for "exciting" states that promise to yield a lot of new knowledge. This topic will be discussed in more detail in the section on 4.3.3 "Artificial Curiosity."

In the next section, we will address the question of how we evaluate the observations we make during exploration. The learning processes of the agents often depend on a whole set of parameters or—e.g., in the case of neural networks (cf. Chap.

5)—also on structural concepts. It was not claimed to present optimal settings. The reader may succeed in improving the examples considerably by various adjustments.

4.2   Retroactive Processing of Experience ("Model-Free Reinforcement Learning")

4.2.1   Goal-Oriented Learning ("Value-Based")

Subsequent evaluation of complete episodes ("Monte Carlo" Method)

What happens when we explore? When exploring an environment, we observe rewards and adjust our estimator function for state evaluations. The so-called "Monte Carlo methods" explore an unknown environmental system in a fairly obvious way. The basic idea in MC studies is to numerically solve complex probabilistic problems using a large number of random experiments. They originate from stochastics and were already studied in the 1940s by prominent figures such as John von Neumann. An illustrative example for the application of the method is the approximation of the circle number π by "sprinkling" a square over a unit circle with random points. From the ratio of points within the circle and the points in total, the circle number π can be determined.

We recall that for each policy π, there is an expected cumulative reward $V^\pi(s)$ that the agent would receive if he used the tactic π from the state s followed. To estimate the value of a state based on a particular policy, we need to compute the expected value in terms of the discounted and cumulative rewards of the subsequent states.

$$V^\pi\left(s_t\right) = E\left[\sum_{i=1}^{\infty}\gamma^{i-1}r_{t+i}\right] = E\left[r_{t+1} + \gamma r_{t+2} + \gamma^2 r_{t+3} + \ldots\right]\left(\text{cf.Chap.2}\right)$$

Monte Carlo methods for our purposes do this by running a large number of episodes and adjusting the value function after observing the result. Monte Carlo algorithms thus determine the rewards of a complete episode and then perform an update of the value function $V^\pi(s_t)$. MC methods are therefore only suitable for episodic scenarios. The state key in the Hamster Gridworlds is represented by a string constructed from the two coordinates of the corresponding tile and a value for the collected grains. It is then used to access the tables (HashMaps) that contain, e.g., the state score $V(s)$ are contained. If the tables in question do not contain an entry for the new key, a new entry is created with the initial values. It is therefore important to always form the state key using this function when learning.

```
public String getStateKey(int x, int y, int score){
  String key="["+x+","+y+","+score+"]"; if(!V.containsKey(key)){
    V.put(key, 0.0); }
  return key;
}
```

Now, to meaningfully estimate the value of a state, we need to average the rewards received over a sufficiently large set of episodes that start in that state. The occurrence of state s of an episode is called the "visit" in s.

However, the same state can occur several times in an episode. In the "first-visit" procedure, the rating is updated only for the first visit in state s.

First-Visit Monte Carlo Prediction, for Estimating V^π (Sutton and Barto)

```
public String getStateKey(int x, int y, int score){
  String key="["+x+","+y+","+score+"]"; if(!V.containsKey(key)){
      V.put(key, 0.0); }
return key;

}
```

The algorithm starts discounting the accumulation of rewards at the end of each episode. Only if a state does not occur again at any earlier point in the episode, its score is updated. The score is updated by taking the new average over the feedbacks received up to that point.

The Monte Carlo Update

A formula that adds another value G to a given arithmetic mean v can be written ad hoc with

$$v := \frac{v \cdot (N-1) + G}{N}$$

In doing so, we restore with $v \bullet (N-1)$; we first restore the absolute sum, add the new value, and finally divide by the new total number of values. The expression can be transformed to

$$v := v + \frac{G-v}{N} \tag{4.4}$$

So we need a statistic $N(s)$ about the number of confirmations available in each case. Together with $\eta = 1/N(s_t)$ and $v = V^\pi(s_t)$ and $R_t = G$, we obtain from the formula Gl. 4.4, the relevant Monte Carlo update rule:

$$V^p(s_t) := V^p(s_t) + h\left[R_t - V^p(s_t)\right] \tag{4.5}$$

R_t describes the respective cumulative and discounted average rewards that accrue if, from time t in state s, policy π is followed. Characteristic for all Monte Carlo methods is therefore that, in contrast to the TD algorithms presented in the next section, the updates are only carried out after each completed episode and that they therefore require scenarios that are episodic.

Javabeispiel MC Hamster
You can examine a Java implementation of the method in the scenario "HamsterWithMonteCarlo," which is stored in the folder "Chapter 4 Decision-Making and Learning." The central code section is found in the "act method," which determines the activities of the agent in the Greenfoot environment. The sequence shown is executed in each simulation step of the Greenfoot environment, i.e., within a repetition loop.

"act" Method of the Hamster Agent in the "MC_Hamster" Scenario

```
public void act(){
 if (s_new==null) {
    s = getState(); }else{
    s=s_new; }
 // apply policy
 double[] P = P_Policy(s);
 int a = selectAccordingToDistribution(P);
 // apply transition model (consider uncertainties) int dir =
transitUncertainty(a);
 // execute action a
 if (dir<4){
    try{
        setDirectionOfView(dir);
        goAhead();
    }catch (Exception e){
        //System.out.println("bump!");
    }
        cnt_steps++; }
 // get new state s_new = getState();
 // get the reward from the environment double r = env.
getReward(s_new);
sum_reward+=r;
 // log experiences episode.add(new Experience(s,a,r));
 if (this.evaluationPhase)
    env.putTracemarker(getSX(s),getSY(s),a,1.0);
 // episode end reached? boolean episodeEnd = false; if
((env.isTerminal(s_new))||(cnt_steps>=max_steps)) {
    episodeEnd = true;
    update(episode);
    startNewEpisode(); }
            if              ((env.DISPLAY_UPDATE)&&(cnt_steps%env.
DISPLAY_UPDATE_INTERVAL==0))
    env.updateDisplay(this);
}
```

It is instructive to let the simulation run slowly at first. One can observe how the hamster initially searches around completely disoriented. If by chance it finds a grain (reward +1) or a trap (reward −1), a new episode is started, i.e., the grains and traps here mark final states. After the hamster has found a grain, the discovered goal-directed episode is quickly optimized. When it has acted according to its policy by the end of the episode, the "Monte Carlo" processing of the observations takes place. The computation of the mean value of the feedbacks is here not realized with a list, as in the shown pseudo code of the algorithm from Sutton and Barto (2018) but is solved with the help of an update calculation, which corresponds to the equation Gl. 4.4. The Monte Carlo update of the state estimation can be implemented in Java as follows:

"Monte Carlo" Evaluation of Episodes in Java

```
protected void update( LinkedList <Experience> episode ){ dou-
ble G = 0;
  while (!episode.isEmpty()){
      Experience e = episode.removeLast();
      String s_e = e.getS();
      G=GAMMA*G+e.getR();
      if (!contains_s(episode,s_e)){
          double avG = V.get(s_e);
          int numGs = incN(s_e);
          avG = avG+(G-avG)/numGs; //calculate new average
          V.put(s_e,avG);
      }
  }
}
```

Although the agent can fully exploit the information of a successful episode in the Monte Carlo procedure, it is also the case that preliminary "discoveries" are particularly strongly "memorized." This can result in the learning process getting stuck in a local minimum. In the simulation, the frequently visited states are marked dark. An exemplary course is shown by ⊙ Fig. 4.1.

The housing arena is not at all easy to tackle for reinforcement learning algorithms because of the narrow doors. In particular, for Monte Carlo algorithms, the algorithm may fail to find its way out of a local minimum in the presence of adverse experiences. Then it can be observed how the paths to the first discovered rewards first optimize and then stabilize more and more or the hamster does not dare to go through the first "door" anymore because the "bottleneck" condition was evaluated strongly negatively after the hamster had fallen into a trap in a previous episode. In this case it is advisable to start with a high exploration parameter, which is then continuously reduced. However, this does not guarantee success. In our case, for example, it can lead to the hamster stepping on traps too often (e.g., near the treasure in the upper left corner), which in turn can have an unfavorable effect on the entire learning process. In the example shown here, the exploration start value was reduced to $\varepsilon = 0.5$ to provoke the described problematic behavior (Fig. 4.2).

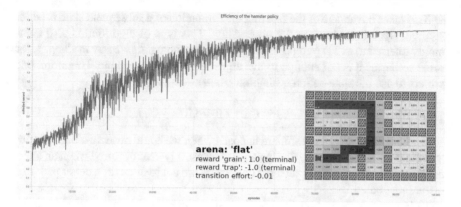

Fig. 4.1 Course of a Monte Carlo evaluation in a gridworld. The darker a tile, the more often the field was visited. The arrows show the course without exploration. An exploration parameter ε melting down to 0 was used. (Parameter: start-ε = 0.7 γ = 0.999 (100,000 episodes)

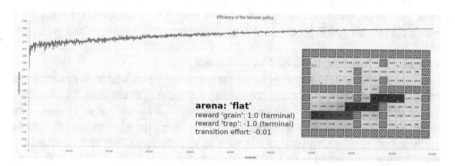

Fig. 4.2 Problematic course of a Monte Carlo evaluation in a gridworld. Again, we used an exploration parameter ε that melts down to 0 but with the slightly lower starting value of 0.5, to elicit the behavior shown (parameter: start-ε = 0.5 γ = 0.999 (100,000 episodes)

Immediate Valuation Using the Temporal Difference (Q- and SARSA Algorithm)

The algorithms discussed in this section apply the insight from Bellman's equation that the value of a state results directly from the evaluations of the subsequent states weighted by the respective transition probability plus the rewards received there or, in the deterministic case, from the one best subsequent state:

$$V(s_t) = \max_{a_t} E\left[r_{t+1} + \gamma V^*(s_{t+1})\right]$$

TD Learning in Deterministic Environments

When exploring, it happens that for a time t, the calculated value $r_{t+1} + \gamma V(s_{t+1})$, which was determined with the observations at time t+1, improves the estimate given until then. In this case, we can adjust the stored valuation accordingly:

$$\hat{V}(s_t) \leftarrow r_{t+1} + \gamma \hat{V}(s_{t+1}) \tag{4.6}$$

This takes advantage of the fact that the estimation of a subsequent state is better secured, since it is closer to the observation. This type of algorithm is also commonly referred to as TD (temporal difference) algorithms. As we saw in Chap. 3, for some scenarios it is not enough to just adjust the $V(s)$ values alone. Therefore, one prefers to use the state-action evaluation $Q(s, a)$:

$$\hat{Q}\left(s_t, a_t\right) \leftarrow r_{t+1} + \gamma \max_{a_{t+1}} \hat{Q}\left(s_{t+1}, a_{t+1}\right) \tag{4.7}$$

Such a Q-update of the estimate happens when we could observe a higher valuation in the subsequent state, for instance, because a reward occurred when an alternative action was randomly selected or because we could discover a better path to a rewarding goal.

● Figure 4.3 shows how the Q-values are adjusted and increase steadily in the process ($\gamma = 0, 9$). If initially only path A is known to the agent, then the transition with the color-coded valuations would have the Q-value $Q(s, a) = 72, 9$, because of $\gamma \bullet max\,(81; 0; 0; 0) = 72, 9$.

If path B is discovered later, then the valuation is changed from $Q(s, a)$ is changed from 72.9 to 90, since the calculation now $\gamma \bullet max\,(81; 100; 0; 0) = 90$ results. As a result, the shorter path B is now used in state s for the agent's future action valuation.

Learning in Nondeterministic Environments

Often, in practice, the outcome of an action is not 100% predictable, for example, in robotics, where movements are sometimes not exactly executable, or in games where chance plays a role, such as backgammon. Due to our lack of control in an uncertain environment, we can only remember probabilities of what reward r can be collected when we choose action a in state s. Moreover, we may also be unable to specify exactly which state s_{t+1} will be reached; if we execute the action a in the state s_t, we execute the action a. By means of a learning rate η with $0 < \eta \leq 1$ which is also called "step size" (often the letter α is often used), we can design the adaptations in such a way that the various possible subsequent evaluations are determined according to their probability distributions; $p(r_{t+1}| s_t, a_t)$ and $P(s_{t+1}| s_t, a_t)$ from the Markov model are taken into account. To achieve stabilization of the evaluations, η can be progressively decayed during the learning process. For this purpose, visit statistics can be used as already described. $N(s, a)$ can be used. If we decrease

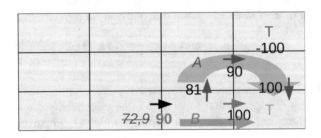

Fig. 4.3 Q-update of a state action evaluation

the learning rate η with the number of visits, then we also force a convergence of the learning process in this way.

This leads us to the famous Q-learning algorithm from Chris Watkins (1989), one of the first crucial breakthroughs in reinforcement learning. Central to this is the update of the Q-function according to the difference between the valuation estimate given so far and the valuation obtained from the scaled estimate of the subsequent state:

$$\hat{Q}(s_t,a_t) \leftarrow \hat{Q}(s_t,a_t) + h\left[r + g\max_{a_{t+1}}\hat{Q}(s_{t+1},a_{t+1}) - Q(s_t,a_t)\right]$$
$$\left(\text{"Q-Learning," Watkins 1989}\right) \hspace{3cm} \text{(Gl. 4.8)}$$

It is also possible again to use only the $V(s)$-values instead of the $Q(s,a)$-values:

$$V(s_t) \leftarrow V(s_t) + h\left[r_{t+1} + gV(s_{t+1}) - V(s_t)\right]$$
$$\left(\text{"TD-Learning," Sutton 1988}\right) \hspace{3cm} \text{(Gl. 4.9)}$$

In contrast to the Monte Carlo approach, we can thus perform "online learning" during the life cycle of the agent and do not have to wait for the end of episodes. Because Q-learning does not require a model of the environment, this method is classified as a "model-free method" of machine learning. For use cases with very long episodes, this waiting for a terminal state can be problematic. Other scenarios are continuous forever and have no episodes at all. In addition, TD methods are also less susceptible to the problem of local minima, since they learn regardless of what subsequent actions are performed in each case.

Q-Learning Algorithm

```
1 Initialize Q(s,a) (arbitrarily f.e. 0, but Q(s_terminal,a) has to be 0 )
2 Loop for each episode:
3     Initialize start s
4     Repeat
5           Choose a from s using policy derived from Q (e.g., ε-greedy)
6           Take action a and observe r and s'
7           Q(s,a) ← Q(s,a) + η[r + γ max Q(s',a') − Q(s,a)]
                                        a'
8           s ← s'
9Until s is a terminal state
```

The deterministic cases can be regarded as special cases of the Q-learning rules, for which the learning rate $\eta = 1$ has been set. The learning rate should be small with $\varepsilon > 0$ and should theoretically be chosen so that the Q-values approximate as closely as possible the average of the discounted and cumulative rewards received in the episodes.

In Q-learning, the system checks whether the valuation of the old state s can be improved. s' whether the evaluation of the old state s can be improved. This is done in a "goal-oriented" and "greedy" manner and thus independent of the actual agent control used, which is partly exploratory. It is therefore said to be an "off-policy" algorithm. Such an agent tends to chase the best of all possibilities and does not consider in its evaluation, for example, dangerous transitions to the same extent as the more realistic "on-policy" algorithms.

"On-Policy" Version of Q-Learning: SARSA Algorithm

The SARSA algorithm uses what is known as "on-policy control." The term SARSA stands for state-action-reward-state-action. While Q-learning always uses the best Q-value for the new observed state, SARSA uses for the update the Q-value that was actually realized, i.e., the one that was selected by a policy that also behaves exploratively and was finally implemented with given transition uncertainties. SARSA thus observes the actual subsequent state, while Q-learning checks all possible subsequent states. On-policy methods estimate the evaluation using the same policy they use to control the agent. Off-policy methods effectively separate the functions. Here, there is one function for state estimation and another function for agent behavior. An agent with Q-learning evaluates "greedy," on the one hand, but, on the other hand, chooses actions, e.g., per ε-greedy, to ensure exploration.

For a greedy agent that does not explore, i.e., which always performs the action with the best Q-value, SARSA and Q-learning are identical. However, if exploration occurs, then they can differ significantly. Because Q-learning always uses the best stored Q-value, the learning algorithm does not care about the current policy that the agent is following. However, Q-learning is more flexible than SARSA as an off-policy, in the sense that a Q-learning agent can learn a good policy even if the optimal strategy is strongly determined by random or uninfluenceable factors. On the other hand, SARSA is more realistic: for some scenarios, it makes sense to assume when learning not only which experience was the best, but what will actually happen in a state with a given control partly determined by chance or other factors? For example, running a red light would still not be advisable even after a possible "sense of achievement."

Sarsa-Learning Algorithm

```
1 Initialize Q(s,a) (arbitrarily f.e. 0, but Q(s_terminal, a) has to be 0 )
2 Loop for each episode:
3     Initialize start s
4     Choose a from s using policy derived from Q (e.g., ε-greedy)
5     Repeat
6         Take action a and observe r and s'
7         Choose a' from s' using policy derived from Q (e.g., ε-greedy)
8         Q(s,a) ← Q(s,a) + η[r + γQ(s',a') − Q(s,a)]
9         s ← s', a ← a'
10    Until s is a terminal state
```

TD algorithms have greatly influenced basic research on artificial intelligence because of their relatively large successes. What is remarkable here is that the algorithms act meaningfully without relying on preconceived knowledge and without constructing an explicit model of the environment. Nor is it a "distributed representation" of the external world. Here it becomes clear that it was not only the so-called "artificial neural networks" that caused this change in viewpoint, but the realization that action is prior to knowledge. With the resurgence of "model-free" machine learning techniques, some researchers argued that it was possible to dispense with explicit representations of knowledge (see also Chap. 6). This has raised debates with proponents of "GOFAI": Is it better to build a model of the environment and a utility function, rather than directly learning an action utility function without a model? What is the best way to represent the agent function (Russell and Norvig 2010, Section 21.3.2.)? Currently, one has left the stage of trench warfare and tries to combine the approaches. In doing so, one realizes that "models" of the environment can make model-free learning more efficient because they allow "virtual discoveries" that are much cheaper than "real" experiments (cf. ▶ Sect. 4.3). In disputes of this kind, it often seems that the solution is to be sought in a synthesis of apparently opposing views.

Java Examples

"Even a blind hamster can find a grain of corn."

In the folder "Chapter 4 Decision-Making and Learning," you can find the scenario "HamsterWithTDLearning." By pressing the "Run" button, the simulation loop is started again. In our case, the behavior of the hamster agent while performing the Q-Learning exploration is similar to that of a blind man groping around in an unknown environment and increasingly learning to follow goal-directed paths. The hamster has nothing more than the two coordinates of the field in which it is currently located. Beyond that, it receives no further sensory information from us. In the scenario, no visit statistics have been implemented, which, according to the concept of "artificial curiosity," could induce the hamster to choose those states that have not yet been visited very often.

One can observe how the evaluations propagate across the state space, starting from the reward states. The rate of propagation is initially small. However, it increases with the number of valued states. Markers are provided at the edges of the boxes, which turn red or blue according to the rating of each action. (Further illustrations on the page https://www.facebook.com/ReinforcementLearningJava; 7.5.2020) (Fig. 4.4).

Q-learning can also be tried with children or youngsters, e.g., in the context of school lessons or a computer science workshop, but in doing so it is recommended, at least at the beginning, to use a deterministic and state-based TD-learning scenario based on Gl. 4.9, at least in the beginning. This works sufficiently well and the principle of the learning algorithm remains easy to understand.

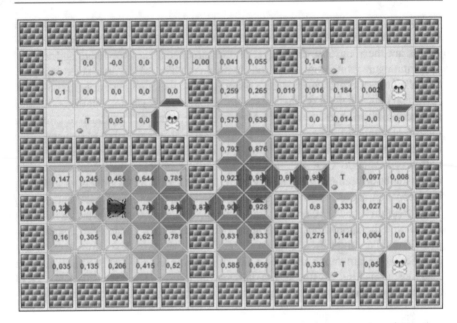

Fig. 4.4 Hamster agent during Q learning in a GridWorld. The numbers in the boxes represent the maximum Q-values. The colored markers at the edges of the boxes represent the current action scores: blue corresponds to a negative score and red to a positive score. Transparency indicates a value close to zero

Relatively quickly, the agent finds the optimal path to the first detected target. Although it is random which grain is found first, a smaller distance to the start state increases the probability of discovery. Feel free to experiment with the learning parameters! You can find them in the upper part of the "Hamster classes." Basically, the TD hamster behaves less "conservatively" than the Monte Carlo hamster and discovers new treasures somewhat more often even after it had already found them elsewhere. However, the algorithm has difficulties to propagate the discoveries quickly. So the agent sometimes cannot benefit from the observations if they are too far away.

Q-Update at Java Hamster

```
protected void update( String s_key, int a, double reward,
    String s_new_key, boolean end ){
  double observation = 0.0; if (end) {
    observation = reward; } else {
    observation = reward + (GAMMA * maxQ(s_new_key)); }
  double q = getQ(s_key, a); q = q + ETA * (observation - q
  setQ(s_key,a, q);
}
```

With "right click" on the blue hamster agent, it can also be removed and replaced by a green SARSA hamster. You can also make this change in the class TD_AgentEnv, then it remains after a reset. For practice purposes, it is recommended to program your own SARSA hamster by manually changing the QHamster to a SARSA hamster.

Conversion of the Q-Hamster into a SARSA Hamster

To do this, create a new subclass "MySarsaHamster" by right-clicking on the corresponding field in the class diagram of the QHamster class. You can also select a hamster image with a different color. For this you have to create a constructor and set the new image file with the Greenfoot command "setImage."

```
public MySarsaHamster(){
    super();
    setImage("hamster_green.png");
}
```

If you now change the line hamster = new QHamster(); in the class TD_AgentEnv to hamster = new MySarsaHamster();, a hamster should appear in the color of your choice. However, this should still behave exactly like the blue QHamster. The update of the Q-table takes place with the QHamster in the function updateQ. You now need a corresponding function updateSarsa and some adjustments in the act method.

Now modify your hamster to follow the SARSA algorithm. Tips: first, copy the commands of the public void act() function from the QHamster completely into the act method of the SARSA Hamster. Most of the algorithms are identical.

The following describes how an implementation of the SARSA Hamster looks like: in SARSA, we not only have to observe the state reached on an action, but we also still have to select the follow-up action using the given policy and the current learning state. Afterwards, the "Sarsa update" is performed; by the way, you can clearly see from the function call update(s,a,r,s',a', end) how the name of the procedure came about.

In the Sarsa update, the new observation is not evaluated with the maxQ(s_new) function, but with getQ(s_new, a_new), taking into account the subsequent action. So the key difference between the Sarsa and Q-update is that instead of reading the maximum Q value, we read the Q value of the subsequent state-action pair that we create with the agent policy. So we are not looking at the "best value"; we are watching what will happen next as a consequence of our action choice.

Sarsa-update in Java

```
protected void update( String s_key, int a, double reward,
    String s_new_key, int a_new, boolean end ){
  double observation = 0.0; if (end) {
```

```
    observation = reward; } else {
      observation = reward + (GAMMA * getQ(s_new_key,a_new) ); }
double q =
getQ(s_key, a); q = q + ETA * (observation - q); setQ(s_key,a, q);
}
```

One scenario in which the difference between Q and SARSA learning is particularly well illustrated is the "cliff walk" (Sutton and Barto 2018). During such a risky walk on the edge of a cliff, it is easy to fall down the cliff due to a strong gust of wind. The shortest path is obviously not the better one in such a case. Since SARSA better accounts for the uncertainties caused by, e.g., exploratory deviations, the "on-policy" algorithm follows a safer path with a larger distance to the "cliff" and thus achieves a much better performance in such a scenario ⦿ (Fig. 4.5).

TicTacToe Agent with Q-Learning

Q-learning can also be used for training board game scenarios. We test the Q-learning procedure in our TicTacToe scenario. Here, the evaluation of field states leading to victory improves in relatively small steps.

We need to adapt our learning procedure to take into account that we have to expect an opponent's move after each of our moves. We take this into account in our update function by negating the expected outcome. The value of a state to us is the opposite of the valuation from the opponent's point of view. Since the rewards are only at the end of the game, i.e., at the end of each episode, the reward does not really matter when evaluating the successor state in this scenario.

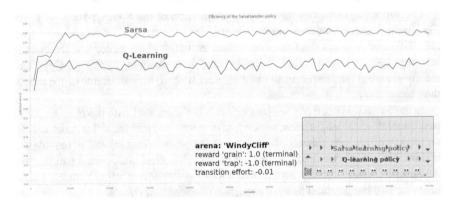

Fig. 4.5 SARSA better accounts for the uncertainties involved in exploratory deviations from the greedy path and thus follows a safer path than Q-learning, resulting in SARSA achieving significantly better performance in such a risky scenario ($\eta = 0.05$, $\varepsilon = 0.05$, $\gamma = 0.999$)

Java Implementation of a Q-Update in a Boardgame Scenario

```
protected void update( int s_key, int a, double reward, int
new_s_key,
boolean end ){
  double observation = 0.0; if (end) {
     observation = reward; } else {
observation = reward + (GAMMA * maxQ(new_s_key));
observation = - observation;    // Invert the valuation, because the
                                // opponent moves. } double q =
getQ(s_key,
a);
q = q + ETA * (observation - q);
setQ(s_key,a, q);
}
```

After initialization of the Greenfoot environment, the training phase starts first when the agent is instantiated. After the specified training intervals, an evaluation phase takes place in which the agent plays as optimally as possible.

You can try out the algorithms in the Greenfoot scenario. We let the algorithm play against itself in the training phase and evaluate first with the optimal playing NegaMax policy. Against the optimal policy, the learning progress is quite uneven. Not much happens over a relatively long period of time until, after a few thousand training games, the successes, i.e., drawn games, come pretty abruptly. By the way, you can also easily copy text outputs in CSV format into a spreadsheet if you want to evaluate them in more detail.

A more even developmental picture emerges when playing against a randomly drawing opponent. The figure shows the learning progress against a randomly playing opponent. The development of playing competence is more clearly visible here.

In TicTacToe, however, there are only on-target moves and wrong moves in a given situation. This means that in this simple game we cannot actually improve on a good move. Therefore, in principle, exploratory behavior is counterproductive in this simple TicTacToe scenario, since there is little gradation in terms of the quality of the moves. We can assume that a goal-directed action is already the best one and in principle do not need to explore further. Therefore, we can also set epsilon to the value 0. Then the algorithm will not lose after only a few hundred games. The ability of Q-learning to discover ever better moves, i.e., shorter paths to the goal, through exploration, i.e., arbitrary "small" deviations from the optimal path, is not needed here. We come up against certain limits of our simple environment (Fig. 4.6).

Consideration of the Action History (Eligibility Traces)

In the examples of TD learning presented so far, only the value of the immediately preceding state is adjusted by means of the temporal difference. We now want to try to extract more information from a completed episode. By carrying an action history, we can adjust the values of the states that were on the path towards a reward.

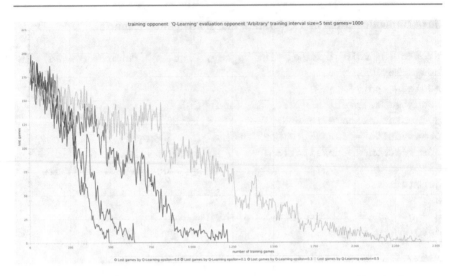

Fig. 4.6 Unfavorable influence of the exploration parameter ε: successes of Q-learning with an ε of 0.5; 0.3; 0.1; and 0.0 against a random TicTacToe opponent

This is done by using eligibility traces, where an additional attribute is added to the state-action pairs $e(s,a)$, which has the value 0 in the "basic state" but gets a value of 1 when the agent enters the state and executes the corresponding action. This identifies which states were on the path that led to the achieved reward. To indicate the decay of the importance of states further in the past for the state score, a decay rate λ is introduced with $0 \leq \lambda \leq 1$. Using $\lambda = 0$ we obtain the algorithms of simple Q or SARSA learning with one-step update; for larger λ values, we obtain a kind of decay rate for the fitness values of the state-action pairs further in the past:

$$e_t(s,a) = \begin{cases} 1, if \ s = s_t \\ \gamma \lambda e_{t-1}(s,a), otherwise \end{cases}$$

So the "suitability" of a "state action" makes a "jump" to one as soon as the agent enters it, and gradually drops thereafter. According to the decay of the suitability $e_t(s,a)$, the learning rate is also reduced. For the SARSA we thus obtain the adaptation rule:

$$Q(s,a) \leftarrow Q(s,a) + e_t(s,a)\left[r_{t+1} + \gamma Q(s_{t+1},a_{t+1}) - Q(s_t,a_t)\right] \ , \forall s,a$$

Because the suitability attributes of *all* state-action pairs must be taken into account, it is now necessary to adjust each of the state-action pairs for the $e_t(s,a) > 0$.

The algorithm obtained in this way is called Sarsa(λ). Analogously, we can also introduce lambda extensions for Q or TD learning. The corresponding algorithms are then called Q(λ) or, more generally, TD(λ).

Sarsa(λ)-learning algorithm

```
1       initialize Q(s,a) arbitrarily,  e(s,a) ← 0 ,∀s,a
2       Loop for all episodes
3            initialize start s
4            Choose a from s using policy derived from Q (e.g., ε-greedy)
5            Repeat
6                 Take action a and observe r and s'
7                 Choose a' using policy derived from Q (e.g., ε-greedy)
8                 δ ← r + γQ(s',a') − Q(s,a)
9                 e(s,a) ← 1
10                for all s,a:
11                     Q(s,a) ← Q(s,a) + ηδe(s,a)
12                     e(s,a) ← γλe(s,a)
13                s ← s',a ← a'
14           Until s is a terminal state
```

Relationship of Monte Carlo to TD(λ)-Learning Methods

Incidentally, the evaluation of complete episodes in Monte Carlo procedures corresponds exactly to the presented TD(λ)-learning procedures for $\lambda=1$, because with this parameter the states are equally adjusted retroactively up to the beginning of the episode. The TD(λ)-procedures, i.e., the procedures that perform learning from the temporal difference with suitability protocols, therefore establish a theoretical link between the "Monte Carlo" sampling, the "TD(1)"-procedures and the single-step exploration, and the "TD(0)"-procedures.

Episodes with high positive or negative returns leave a particularly "lasting impression" on learning procedures that use long paths or entire episodes for state evaluation. While this leads to successful paths being found more quickly because more information is extracted from successful episodes, it also has the negative effect of making it harder to abandon unfavorable paths. TD(0) agents visit all sorts of positive or negative states while roaming around and draw on them for the current state evaluation. In contrast, for a deviation from a successful behavior, Monte Carlo agents require not only leaving the path by exploration behavior or other coincidences at an opportune moment but secondly also successfully completing an alternative episode initiated by it. In the case of the Java hamster in the "flat" arena, it can sometimes be observed that it no longer ventures out of a "room" if it had entered a trap in an early episode. This gave the "door" field a strong negative rating, regardless of the fact that there are many other very rewarding paths behind the door.

Furthermore, unlike MC methods, TD learning methods with $\lambda<1$ can also be used for non-episodic scenarios. Moreover, in TD learning, the feedback is processed directly after each time step, which reduces the complexity of the feedback since the observed result depended on fewer previous transitions and rewards, which usually has a positive effect on the learning success of the algorithm.

Sometimes, instead of a "suitability protocol," a certain number of past states is simply taken into account in the evaluation. These algorithms are then called "n-step" procedures. For example, a five-step SARSA always updates five states. This approach also represents a very interesting compromise between the two extremes. You can also look at this as such an agent always looks ahead a certain number of steps. You can find more information on this in the ▶ section "Technical improvements to the actor-critic architecture" for more information.

Analogous to the two approaches to state evaluation of "value iteration" and "policy iteration" presented in the context of dynamic programming in Chap. 3, the following section will examine how a strategy related to iterative tactic search can be pursued in the exploration of an environmental system. In the former case, starting from discovered goal or reward states, the respective valuations are generated or "updated"; in the latter case, starting from the agent's respective state, we now have to test what consequences a certain action would have. Afterward, however, we do not update the evaluation, which would otherwise only be the well-known tactics-based ("on-policy") state evaluation, but we improve the control directly, i.e., we check whether the assignment of an alternative action option in the respective state would improve the tactics.

4.2.2 Policy Search

Previously, we assumed greedy control in our agents, to which we presented valued successor states. We optimized the agent's behavior by improving these valuations. Now we take a different approach.

The learning process now takes place by improving not the evaluation of the states but directly the assignment of the actions to the states. An explicit representation of the action probabilities directly associated with the states, i.e., without the intermediate step of evaluating the different action options, can reduce the resources needed for action selection. In the tactic search approach, we start with an arbitrary policy and try to optimize it by assigning better actions to the states. A first approach that "reverses" the roles of state-value function and policy in the production process of desired behavior in this sense is the Monte Carlo tactic search approach.

Monte Carlo Tactics Search

In the reinforcement learning methods considered so far, the agent control system π a valuation function $V(s)$ or $Q(s, a)$ is used to evaluate the possible actions in a state. These evaluations correspond to the—more or less well estimated—gain that the agent has when it is in state s and from there selects the "optimal" actions to the target states. In doing so, we either strive as greedily as possible toward the recognized goals independently of our given control ("off-policy") (value iteration, Q-learning) or take into account—"on policy"—our given tactics and evaluate the actions more "realistically" and observe what actually happens (Monte Carlo, SARSA). Nevertheless, both correspond in principle to an approach with goal-oriented state evaluations.

The basic idea of policy search is to adapt a given tactic until its performance no longer improves. What might a tactic search in unknown environments look like, analogous to the policy iteration scheme presented in Chap. 3?

In an unknown environmental system, the agent has no or very few helpful mappings available at the beginning. Accordingly, the agent behaves arbitrarily, ineffectively, and the rewards obtained are very small. To improve the policy, we need to assign better actions to the states so that we get larger rewards.

In the following example, we use a rating function $Q(s, a)$, but we only use it to store the best action for the visited world states in the policy. By using the evaluation function for our search, we obtain a "compass" for promising modifications to the policy. The roles of policy and evaluation function have thus been swapped.

In a given state, the agent must on the one hand perform appropriate actions but, on the other hand, also remain flexible and explore new things from time to time. This means that we save the reward received afterward for an action but, on the other hand, the policy must not exclusively follow the "best" actions. We need "soft" policies, so to speak, that do not make completely deterministic decisions.

The solution is that, with respect to the discovered new valuations $Q(s, a)$, use a probabilistic, i.e., say, an ε-greedy adjustment of the policy. The policy update therefore only changes the policy in the direction of the greedy tactic.

For this purpose, we will present here the algorithm "first-visit MC-Control for ε-soft policies" (Sutton and Barto 2018), an algorithm that combines a Monte Carlo evaluation with a "policy improvement" that works with so-called "soft" ε-policies. For this purpose, the authors introduce the notion of ε-soft policy.

For ε-soft policies, the probability of selecting a particular action never falls below a certain lower bound, i.e., there always exists an out probability of greater than 0 for an action a in a particular state s.

An ε-soft policy is any policy for which it holds that:

$$\mathbf{p}(s, a) = P_{\mathbf{p}}(a|s) \geq \frac{\mathbf{e}}{|A(s)|}$$

Within the set of ε-soft policies, the ε-greedy tactics are those that are closest to a pure greedy function.

In Sutton and Barto (2018, Ch. 5.4), it is shown that an ε-greedy policy π', which uses for action selection a state-action evaluation Q^π of any other ε-soft policy π, is in every case better than or as good as π. This allows us to apply it to policy improvement as we learned about it in the dynamic programming section. Improvement consists of turning an ε-soft policy into an ε-greedy policy.

The estimation of the $Q(s, a)$ state action values is again done according to the "First-Visit-MC-Method," in which the first visit of a state action pair is updated with the average of the subsequently cashed evaluations.

"first-visit Monte Carlo control for ε-soft policies" (Sutton and Barto)

```
1       π ← an arbitrary ε-soft policy
2       Q(s,a) ∈ ℝ (arbitrarily), for all s ∈ S,a ∈ A(s)
3       returns(s,a) ← empty list, for all s ∈ S,a ∈ A(s)
4       repeat for each episode:
5               Generate an episode following π: s₀,a₀,r₁, s₁,a₁,r₂, … ,s_{T-1},a_{T-1},r_T
6               G ← 0
7               Loop for each step of episode, t = T − 1, T − 2,…,0:
8                       G ← γG + r_{t+1}
9                       Unless the pair s_t,a_t appears in s₀,a₀, s₁,a₁,… ,s_{T-1},a_{T-1}:
10                              Append G to returns(s_t,a_t)
11                              Q(s_t,a_t) ← average(returns(s_t,a_t))
12                              a* ← argmax_a Q(s_t,a_t)
```

$$13 \qquad \pi(s_t,a) \leftarrow \begin{cases} \frac{\varepsilon}{|A(s_t)|}+1-\varepsilon & if\ a = a^* \\ \frac{\varepsilon}{|A(s_t)|} & if\ a \neq a^* \end{cases}$$

The algorithm follows successful paths but also leaves room for exploratory deviations.

You can observe the behavior of the algorithm in the scenario "HamsterWithMonteCarlo" in the folder "Chapter 4 Decision-Making and Learning." To do this, you must change the "MC_Hamster" agent to a "MC_PolicySearch_Hamster" in the constructor of the MC_AgentEnvironment class. Due to the possibilities of object orientation, many attributes and functions can be inherited from the MC-Hamster class. It is convenient that in the unclasses only the special features have to be either overloaded or reprogrammed. Policy markers also now appear in the visualization, indicating which actions the hamster favors at each state.

Of particular interest, perhaps, is the section that evaluates the episodes and how to improve the policy.

Monte Carlo policy search for ε-soft policies in Java

```java
protected void update( LinkedList <Experience> episode ){
  double G = 0; while (!episode.isEmpty()){
    Experience e = episode.removeLast();
    String s_e = e.getS();
    int a_e = e.getA();
    G=GAMMA*G+e.getR();
    if (!contains_sa(episode,s_e,a_e)){
        double avG = getQ(s_e,a_e);
        int numGs = incN(s_e,a_e);
```

```
        avG = (avG*(numGs-1)+G)/numGs;
        setQ(s_e,a_e,avG);
        incN(s_e);
        int a_max = getActionWithMaxQ(s_e);
    // epsilon-greedy
        ArrayList <Integer> A_s = env.coursesOfAction(s_e);
        double[] P = new double[SIZE_OF_ACTIONSPACE];
        int k = A_s.size();
        for (int a_i : A_s){
            P[a_i] = current_epsilon/k;
        }
        // For a_max add the probability 1-Epsilon.
        if (a_max>=0) P[a_max]+=(1-current_epsilon);
        // policy update
        setPi(s_e,P);
    }
  }
}
```

In the exemplary course shown here, an exploration parameter melting from 0.7
to 0 was again used. ε was used (Fig. 4.7).

Are there also possibilities to completely dispense with condition $Q(s,a)$ completely? An obvious variant for this is to modify policies purely randomly in a large
population of agents, to test them, and to select the most successful variants in order
to finally submit them to a further test run. This corresponds to the approach taken
by genetic algorithms, i.e., algorithms that perform a kind of "artificial evolution,"
although here we are dealing more with a kind of goal-directed "breeding." We

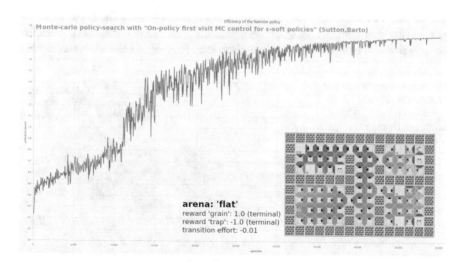

Fig. 4.7 Progression of a Monte Carlo policy search with ε melting down from 0.7 to 0

consider this in the next section. Before that, however, we take a look at properties of natural cognitive systems.

Evolutionary Strategies
Properties of Biological Controls

In Godfrey-Smith (2019), naturalist, philosopher of science, and avid deep-sea diver Peter Godfrey-Smith outlines the evolutionary history of animals, from their early unicellular beginnings some 3.8 billion years ago to the currently observable forms of "vertebrates," "mollusks," "arthropods," "jellyfish," and "sponges."

For most of these 3.8 billion years, the biological world was dominated by single-celled organisms. Sometimes the idea is widespread that such unicellular organisms just passively float around in their milieu. The opposite is evident from the behavior of the widely distributed bacterium *Escherichia coli* ◉ (Fig. 4.8) which plays an important role in the intestines of animals and humans but can also cause diseases.

Under the microscope, it can be observed how *Escherichia coli* actively move. The "motor" part is performed by threadlike structures that are evenly distributed over the cell surface, the "flagella." The bacterium can move in a straight line in one direction by bundling some flagella and working together. Sometimes, however, straight-line locomotion is aborted as the flagellar bundle disintegrates and the individual filaments turn in different directions; the bacterium then begins to tumble. The flagellar bundle eventually reorganizes itself and the bacterium accelerates in a different direction. The bacterium's action set therefore consists of two activities "straight-line movement" and "tumbling." It thus chooses between "going straight" and a random change of direction.

The sensory system of the bacterium consists of collections of molecules that penetrate the cell membrane and, in a sense, establish a connection between the inside and the outside, giving the bacterium a sense of smell and taste, so to speak, that allows it to sense chemical substances in the environment. The bacterium cannot actually determine which direction a harmful or attractive chemical substance is coming from. The trick of the bacterium is to keep moving. It senses whether certain

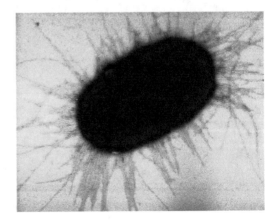

Fig. 4.8 Portrait image of an *Escherichia coli* bacterium (Gross L, Public Library of Science Biology Vol. 4/9/2006, e314. Image: Manu Forero)

concentrations are rising or falling. For example, if it senses that the concentration of an attractive substance is increasing, the probability of selecting the tumbling motion decreases, and the bacterium more often moves purposefully straight ahead.

The tiny bacterium *Escherichia coli* is seen from the outside in a huge unknown environment, e.g., in an intestine. For the bacterium, however, this world is completely inaccessible. Heinz von Foerster put it this way: an observer does not see what he does not see; in other words, as an observer you do not notice that you cannot see many things at all. Therefore, the world of the bacterium consists initially only of the change values of certain chemical substances, which corresponds to a comparatively manageable state space. The problem of the bacterium's policy is to determine under which circumstances it chooses action one "straight-line motion" or action two "random tumbling."

Multicellular biological organisms emerged about 1 to 2 billion years ago. The process is not entirely clear, but bifurcations occurred in biological evolutionary history, at which the "children" of certain evolutionary "links" eventually took different developmental paths, resulting in fundamental directional decisions with regard to further physical development and, correspondingly, with regard to the manner of cognition and the further development of behavioral control. Sometimes new forms of cooperation and higher levels of behavior also developed. Thus, some aggregations of genetically identical single-celled organisms evolved into multicellular units, whereby the once independent organisms began to function as part of larger structures. At this moment, an evolutionary optimization of the cooperation, the "teamwork" of the respective single organisms began, first through the exchange of chemical messengers and then also with the development of specialized organic units also for cognitive tasks such as sensory, motor, or communication, which then also caused the emergence of the first "neurons."

According to Godfrey-Smith, the bifurcation of vertebrates and mollusks took place about 600 million years ago, which means that the common ancestor of man and octopus lived at this time. He sketches this ancestor as a small worm-like creature that already had small eye spots at the tip of its body and an early form of a spinal cord-like nervous system that could conduct signals through the body. On the one hand, the neurons took over sensory-motor coordination, i.e., the purposeful processing of stimuli from the environment, but also tasks of successful "action formation." The two tasks of the cognitive system are reminiscent of the presented duality of "policy-based" and "value-based" action strategies.

Godfrey-Smith points out that the view of "action formation" in general often falls short. The aspect of successful "action formation" is of central importance in the emergence of cooperative ways of life, especially for the emergence of complex multicellular organisms. In the case of complex agents or multi-agent systems, it is important that various active system components can interact meaningfully and expediently in order to be able to produce goal-oriented courses of action at all. To illustrate the problem, imagine a rowing ship such as an ancient trireme. This needs not only a helmsman who keeps an eye on where the ship is heading, i.e., where there are worthwhile targets or dangers, but also the rowers and, for example, a person who, with the help of an acoustic signal generator, sets the stroke so that the individual actions are coordinated. All energy is wasted if, for example, the paddles of a rowing boat simply move wildly without coordination. Certainly, the emergence of language and consciousness can also only be properly understood from the perspective of cooperative action formation. In the context of this book, however, we will have to limit ourselves to the view of individually working "cybernetes" (Greek for "helmsman").

In principle, it is not very plausible that in the early "simple" forms of cognition an explicit evaluation of different options for action is made. Rather, it is to be assumed that given, somehow directly "wired" controls are modified and optimize themselves in the course of time by the fact that those variants reproduce themselves increasingly and prevail which simply work "better." Therefore, in the following we want to take a look at policy search methods that largely refrain from elaborately building an evaluation function. This also means that we refrain from first evaluating the various options for action before making a decision. In the following, we introduce the concept of "stochastic policy," where we can assign certain selection probabilities to the possible actions. In doing so, we also say goodbye to the search for the one "right" decision.

Stochastic Policies
In the tactics improvement so far, we have used evaluation functions to see whether there might be better evaluated actions in a situation and then adjusted the assignment accordingly. Ultimately, it was always a matter of finding the optimal decision for a given situation, from which we then deviated only occasionally for exploration purposes.

In the learning algorithms with stochastic policies, we use parameterized controls that give us a probability distribution over the possible actions $\pi(s,a,\boldsymbol{\theta}) = P(a \mid s, \boldsymbol{\theta})$ with ($a \in A$). We can now adjust this continuously, i.e., without "jumps," by "turning" the "adjusting screws" $\boldsymbol{\theta}$ "turn." We no longer want to find out the optimal action in each case but rather devote ourselves to the question of an optimal probability distribution.

In order to be able to make continuous changes to a stochastic controller, the SoftMax distribution comes in handy. Recall that it allows us to elegantly produce a probability distribution over the various possible actions.

$$\pi\left(s,a,\theta\right) = \frac{e^{h(s,a,\theta)}}{\sum_{b \in A}^{A} e^{h(s,b,\theta)}} \tag{4.10}$$

Since we don't have a "proper" evaluation function, we use "action references" $h(s,a,\boldsymbol{\theta})$ which are generated from the parameters. These preferences do not represent realistic valuations. For example, it may be that—if the optimal tactics require deterministic decisions and the parameterization allows this—the preferences of the optimal actions tend to infinity compared to those of the suboptimal actions. The SoftMax function normalizes these values so that the sum of the probabilities equals the 1 required by definition. At the same time, the ratios of the values in the distribution generated by the function are well represented.

Genetic Algorithms

Genetic algorithms can be used to optimize attributes of individuals with respect to desired characteristics. The idea of genetic algorithms is to evaluate individuals with respect to the desired goal, select them from the population, and propagate them.

This is reminiscent of targeted breeding selection. This method is used, for example, in plant breeding to produce desired characteristics, the central method for this selection and crossing of suitable individuals. The characteristics of the individuals are determined by their genetic material. Biological organisms use their genome for the transmission of hereditary information, which is contained in each individual cell of the organimus. The information needed to inherit traits is contained in DNA and is encoded by sequences of the DNA bases adenine (A), guanine (G), cytosine (C), and thymine (T). Sections on the DNA are also called genes. Genes can code for proteins, for example, which are then formed in the cell.

It should be noted, however, that ontogenetic characteristics, especially in higher life forms, are not deterministically determined by genes, but are formed through a complex and multilayered process. On the level of heredity, for example, gene regulation, i.e., the control of gene activity, also plays an important role. It determines, for example, whether, when, and in what quantity the protein encoded by the gene is formed in the cell. In addition, there are the various adaptation and modification processes in the respective higher stages, which determine which properties and characteristics an organism ultimately develops.

How is each new generation generated in genetic algorithms? In "artificial evolution," too, attributes of individuals are determined by a so-called "genome," whereby a "population" consists of a number of "genomes." In our case, a genome encodes the parameters of the policy. This means that from each genome a certain agent control can be generated.

The usual steps in a genetic algorithm are as follows:

1. initialization of the gene pool: The first generation of genomes is generated randomly.

Go through the following steps until a termination criterion is met:

1. Production: creation of the new generation according to the present gene pool.
2. Evaluation: Each candidate solution of the generation is assigned a fitness value.
3. Selection: selection of individuals for recombination and mutation.
4. Variation: for example, by mutation (random change of descendants) or recombination (mixing of selected individuals).

Implementation of an Evolutionary Optimization

A Java implementation of a genetic algorithm that adjusts stochastic policies in a gridworld can be found in the folder "Chapter 4 Decision-Making and Learning." The scenario is called "HamsterWithEvolutionaryStrategy."

In evolutionary strategies, the environmental class plays a greater role in "learning." This is because no learning actually takes place within the agent, and the adaptations, mutation, interbreeding, selection, and production of the next generation are basically done "from the outside." A convenient entry point for studying the algorithm in this case would be the code of the World class "Evolutionary_PolicySearch_Environment."

The genomes of a generation are held in the class "GenePool." The class also provides the various methods of the genetic algorithm. The initialization of the gene pool takes place in the constructor of the world class. When the constructor of GenePool is called, the function "generatePool" is called which generates the appropriate number of genomes. In the genomes the assignment table of the policy is mapped. A special feature of the implementation is that the genome is generated "empty" and the entries "genes" for the visited states are added. This allows different maps to be used without first generating genomes for all possible world states. The "genes" are thus generated when new states are visited and initialized with arbitrary values (normal distribution with standard deviation "new_gene_sigma"). The genes are also passed on to the offspring and also distributed in the course of interbreeding.

Therefore, setting the arena can be done as usual at the beginning of the constructor with the call to super, which calls the constructor RL_Evo_GridEnv(String[] fielddescription) of the parent class. Feel free to try a different environment where you set the arena "mapFrozenLake."

The production of the agents according to the current gene pool happens in the function "makePopulation." The function expects the desired start state as argument.

In general, as mentioned earlier, the agents play a relatively passive role in this learning algorithm. They react according to their specified control and collect more or less reward. When an agent finishes an episode, it calls the function "evaluation-Finished." The accumulated reward is now noted as a fitness value at the corresponding genome and used for selection. After that, it checks whether agents of the current generation are still active. If no more agents are active, the selection and varying reproduction (crossing, mutation) of the genomes of the next generation is performed by calling the function "genePool.breedNextGeneration."

Taking the Fitness Value After End of an Episode and Start Breeding Next Generation

```
public void evaluationFinished(Genome genome, double fitness){
genePool.setFitness(genome.ID,fitness);
  living_agents--;
  if (living_agents<=0){
     double avg_fitness_topgroup = genePool.breedNextGeneration();
     makePopulation(getHamsterStartX(),getHamsterStartY());
updateDisplay(genePool.getTopGroup());
        jfxLogger.append(genePool.getGeneration(),avg_fitness_top-
group); }}
```

The breeding of the next generation is carried out as follows: First the top group is determined. This is completely carried over into the next generation, which makes regression in the development of fitness more difficult. Then the individuals of the top group are crossed each with each other. This produces two complementary individuals each, both of which are carried over. Thereby $n(n - 1)$ hybrids are added to the top group of the next generation. The remaining population is filled with mutant clones of the top group (75%) and random individuals (25%).

The method presented here, which performs the breeding of the next generation, does not claim to be particularly efficient and has a strongly intuitive character, e.g., the way the crossing of the genomes is performed. In implementation, the genes of the parents are randomly selected in equal proportions. Often this is done differently, e.g., by splitting the genomes, from A_1A_2 and B_1B_2, e.g., the "children" A_1B_2 and B_1A_2. In practice, it is often still the case that much depends on the intuition of the developer. Experiment with your own modifications!

Method Used to Breed the Next Generation (Class 'GenePool')

```
public double breedNextGeneration(){
 bestGenomes = getBestGenomes(num_top);
 ArrayList<Genome> children = new ArrayList<Genome>();
 // take over top group to next generation
 double avg_top_fitness=0;
 for (Genome gene : bestGenomes){
     avg_top_fitness+=gene.fitness;
     gene.fitness=0.0;
     children.add(gene);
 }
 avg_top_fitness/=bestGenomes.size();
 // mix up top group individuals
 for (int i=0;i<bestGenomes.size();i++){
     for (int j=i+1;j<bestGenomes.size();j++){
         Genome[] mixed = mix_it(bestGenomes.get(i),bestGenomes.
get(j));
         children.add(mixed[0]);
         children.add(mixed[1]);
     }
 }
 // fill up remaining population with mutated clones of the top
group (75%)
and random individuals
 int remainingChildren = (env.POPULATION_SIZE - children.size());
 for (int i = 0; i < 3*remainingChildren/4+1; i++) {
     children.add(clone_mutate(bestGenomes.get(i%num_top)));
 }
 for (int i = children.size(); i < env.POPULATION_SIZE; i++){
```

```
    children.add(new Genome());
}
clearPool();
for (Genome next_gen : children) put(next_gen);
cnt_generation++;
return avg_top_fitness;
}
```

Stochastic policies allow "small" mutations. This results in a more continuous and uniform learning process. The development process is also more stable if the population and the selected top group are extensive.

The results in the simulation runs show an astonishingly fast adaptation considering the rather "arbitrary" and highly random method. Often a good policy is found for the small environments after relatively few generations. However, "spontaneous" fitness collapses can occur when unfavorable circumstances cause a negative mutation to spread throughout the entire top group. This occurs especially in "risky" environments such as the so-called "frozen lake" environment (https://gym.openai.com/envs/FrozenLake-v0/; 05-08-2021). Here, a frisbee must be reached by the agent on an unstable ice sheet (Fig. 4.9).

The progress of adaptation is determined by a larger number of parameters. These are not only the size of the top group, the type, and extent of mutations and crosses. Unfortunately, the properties and effects of the parameters cannot be discussed in more detail within the scope of this book, the purpose of which is to provide an overview of how the various approaches work. Experiment with different settings and observe the effects on the adaptation process. Sharing results can lead to interesting exchanges. The following parameters were used in the runs shown:

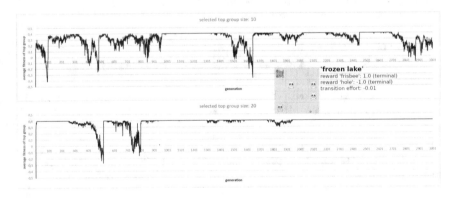

Fig. 4.9 Average fitness (cumulative reward) of the top group in the "frozen lake" scenario as a function of generations. The total population was 1000 individuals. Instabilities occur more when only a smaller group is selected

parameter in 'World' class 'Evolutionary_PolicySearch_Environment'

```
/* size of the gene pool */public
static final int POPULATION_SIZE = 1000;
/* size of the top group */public
static final int SIZE_OF_TOPGROUP = 20;
/* determines percentage of genes (entries of the policy table)
that will
be arbitrary manipulated */public
static final float MUTATION_RATE = 0.9f;
/* standard deviation of the modifications */public
static final float MUTATION_STANDARD_DEVIATION = 0.005f;
```

In addition, properties of the policy, e.g., the temperature parameter T of the SoftMax function or the evaluation (e.g., the number of repetitions for creating an average) also play a role.

parameter in 'Actor' Class 'Evolving_Hamster'

```
public static double NEW_GENE_SIGMA = 1; // standard devia-
tion, if new
random genes are produced
public final static double T = 1; // exploration parameter in softmax
("temperature")
public final static int EVALUATION_INTERVAL = 3; // interval for
making an
average and displaying a result.
```

The amount of computing time required can be enormous, especially for large populations and small mutation rates and for larger state and action spaces. Purely random modifications require a lot of computing power, and the results are not very transparent, so in larger state and action spaces, assistance must usually be built in. These can be, e.g., "roads," which punish the deviation from an externally determined baseline, or the establishment of continuous or frequent rewards, e.g., when the agent reaches milestones.

With a population size of 1000 and 3 evaluation repetitions, 3000 evaluation episodes must be performed in every single generation. It is obvious to make the parameters themselves the subject of a learning and optimization algorithm. The keyword for this is "meta-learning."

The approach of evolutionary algorithms has some similarities to the posterior evaluation of complete episodes in the Monte Carlo approach. However, in evolutionary algorithms we completely avoid building a state estimation function, which saves some resources, but useful experiences are only perpetuated by selecting the most successful examples (genomes) and discarding the less good ones.

In genetic algorithms, the agents are sifted out via the selection process, i.e., they are passive like small stones. In reinforcement learning, on the other hand, one is very interested in individual learning processes within active agents that use "experiences" for a purpose. Is it possible to pursue the policy search approach and give the individual agents a more active role in the learning process?

Monte Carlo Policy Gradient (REINFORCE)

The question that will concern us in what follows is can the "adjusting screws" θ of the agent policy not only arbitrarily but purposefully "turned" by processing individual experiences gathered by the agent in an episode? The algorithms presented in the following essentially refer to the explanations of Sutton and Barto (2018).

For this purpose, we introduce a performance measure $J(\theta)$ which is supposed to give us clues about how good our control is depending on θ and whether our agent control is improved or worsened by the changes in the parameters. θ improves or worsens.

$J(\theta)$ creates a (hyper-)plane over the parameters θ; the "higher" the value $J(\theta)$, the better the performance. We now try to design the learning in such a way that we maximize the performance measure by following a gradient ascent:

$$\theta_{t+1} = \theta_t + \eta \nabla J\left(\theta_t\right) \tag{4.11}$$

An important advantage of a "soft" parameterization is that it allows the setting of arbitrary action probabilities. Methods that make "hard" decisions according to action values offer no possibility to find stochastic optima. If, on the other hand, we continuously approximate policies, we can also form controls with a specific probability distribution that more elegantly represents the risks and potential rewards in an environment.

Now, how can we modify the policy parameters to ensure performance improvement? First, for episodic cases, for example, we can identify the performance of our controller with the value of the initial state under a given controller π_θ identify:

$$J\left(q\right) = V^{P_q}\left(s_0\right)$$

With which also a corresponding performance gradient can be represented:

$$\nabla J\left(q\right) = \nabla V^{P_q}\left(s_0\right)$$

Theoretically, however, we now get a problem which consists in the fact that a given performance depends not only on the choice of actions but also on the distribution of future states in the environment in which this choice is made by the corresponding policy. The "performance gradient" is thus not only dependent on the policy but also represents a property of the environmental system in which the agent operates. How can we estimate a performance gradient that depends only on policy parameters when this gradient depends on an unknown effect of these changes on the seen state distribution?

This problem is solved with the so-called "policy gradient theorem," which can be formally expressed with:

$$\nabla J(\theta) \propto \sum_s \mu(s) \sum_a Q^\pi(s,a) \nabla \pi(a|s,\theta) = E_\pi \left[\sum_a Q^\pi(s_t,a) \nabla \pi(a|s_t,\theta) \right] \quad (4.12)$$

(noting Sutton and Barto 2018) By $\pi(a|s_t, \theta)$ is here in principle $P_\pi(a|s_t, \theta)$ is meant. The right-hand side of the theorem represents the weighted sum over the expected discounted rewards as they would occur under policy π if it is followed accordingly. The formula states that our "performance gradient" emerges from the expected value over the summed products of state valuation and the "policy gradient." A formal proof of the theorem is provided by Sutton and Barto (2018). In principle, this means that if we want to follow a performance gradient, we have to increase the selection probability of those actions that promise a large reward.

How can we translate this theory into algorithms? Although the last term with the expected value E_π contains concrete sequences of states s_t, to compute the expression Gl. 4.12, we would have to look at all the actions $a \in A(s_t)$ in order to calculate this sum.

We therefore reshape the expression in such a way that we can calculate the gradient from the action chosen at each time t a_t at the time t:

$$\nabla J(q) = E_p \left[\sum_a Q^p(s_t,a) \nabla p(a|s_t,q) \right] = E_p \left[\sum_a Q^p(s_t,a) p(a|s_t,q) \frac{\nabla p(a|s_t,q)}{p(a|s_t,q)} \right]$$

$$= E_p \left[Q^p(s_t,a_t) \frac{\nabla p(a_t|s_t,q)}{p(a_t|s_t,q)} \right] \quad (4.13)$$

With the consideration of the concrete actions a_t of a sample, which are selected in large numbers according to the probability $\pi(a_t|s_t, \theta)$, we omit the "weighted" sum over all actions.

Monte Carlo Policy Gradient (REINFORCE)

For the members (s_t, a_t) of a sample of length T, the respective "measured" valuation results from the discounted and cumulated subsequent rewards (cf. ▶ Chap. 2):

$$Q^\pi(s_t,a_t) = \sum_{k=t+1}^{T} \gamma^{k-t-1} r_k = G_t$$

From Gl. 4.13 we can use this information to establish the update rule of the classical REINFORCE algorithm (Williams 1992) set up:

$$\theta_{t+1} = \theta_t + G_t \frac{\nabla \pi(a_t|s_t,\theta)}{\pi(a_t|s_t,\theta)}$$

With the transformation rule $\nabla \ln x = \frac{\nabla x}{x}$, we get

$$\theta_{t+1} = \theta_t + \eta G_t \nabla \ln \pi(a_t|s_t,\theta) \quad (4.14)$$

where $\eta > 0$ again represents a learning parameter that controls the step size and thus the learning speed but also the "granularity" of the process.

Monte Carlo Policy Gradient Algorithm REINFORCE

1   π_θ a differentiable stochastic policy parameterization $\pi_\theta(a,s,\boldsymbol{\theta})$

2      $\eta_\theta > 0$ parameter for the adjustment of $\boldsymbol{\theta}$ (step size)

3   Initialize policy parameter $\boldsymbol{\theta} \in \mathbb{R}^d$

4   Loop for each episode:

5   Generate an episode π_θ: $s_0, a_0, r_1,\ s_1, a_1, r_2,\ \dots, s_{T-1}, a_{T-1}, r_T$

6   Loop for each step of the episode $t = 0,1,\dots,T-1$:

7               $G \leftarrow \sum_{k=t+1}^{T} \gamma^{k-t-1} r_k$

8               $\boldsymbol{\theta} \leftarrow \boldsymbol{\theta} + \eta_\theta\, G_t\, \nabla \ln \pi(a_t, s_t, \boldsymbol{\theta})$

This already makes a very compact and practical impression, but the question is still open how the "policy gradient," i.e., the term $\nabla \ln \pi(a_t | s_t, \boldsymbol{\theta})$, is calculated. We recall that for our parameterized policy, we have the probability distribution P_π with:

$$P_\pi\left(a|s,\theta\right) = \pi\left(s,a,\theta\right) = \frac{e^{h(s,a,\theta)}}{\sum_{b\in A}^{A} e^{h(s,b,\theta)}} \tag{4.15}$$

want to calculate, i.e., a "SoftMax" selection strategy with the respective action references $h(s,a,\boldsymbol{\theta}_s)$.

To tie in with what we know, we assign a preference to each state-action pair $\theta_{s,a}$ to each state-action pair. This means we set:

$$h\left(s,a,\theta_s\right) = \theta_{s,a}, \text{where } \theta_{s,a} \in \mathbb{R}$$

which allows us to simplify:

$$\pi\left(s,a,\theta_s\right) = \frac{e^{\theta_{s,a}}}{\sum_{b\in A}^{A} e^{\theta_{s,b}}}$$

respectively

$$\ln \pi\left(s,a,\theta_s\right) = \theta_{s,a} - \ln \sum_{b\in A}^{A} e^{\theta_{s,b}} \tag{4.16}$$

are obtained. For the components of the gradient $\nabla \ln \pi(s,a,\boldsymbol{\theta}_s)$, we now obtain by derivation:

$$\nabla \ln \pi\left(s,a,\theta_s\right) = \frac{\partial \ln \pi\left(s,a,\theta_s\right)}{\partial \theta_{s,i}} = \begin{cases} 1 - \dfrac{e^{\theta_{s,a}}}{\sum_{b\in A}^{A} e^{\theta_{s,b}}} = 1 - \pi\left(s,a,\theta_s\right) & \text{for } i = a \\[4mm] 0 - \dfrac{e^{\theta_{s,i}}}{\sum_{b\in A}^{A} e^{\theta_{s,b}}} = -\pi\left(s,i,\theta_s\right) & \text{for } i \neq a \end{cases}$$

Which we have used to create a policy gradient that we can use to test this theory in our Java environment. We have thus obtained a "tabular" variant of the policy gradient algorithm. For small state action spaces such as in the box worlds, this is, after all, practical. The use of approximators will be discussed in the next chapter. This will then also lead to some problems for which some work-arounds have been developed in the course of applying deep learning.

For our purposes, a clear view of reinforcement learning from scratch, it is convenient that we keep a manageable number of states and that we have a localized update which leaves the action probabilities of the other states untouched.

Java implementation of the REINFORCE update (tabular policy parameterization)

```
protected void update( LinkedList <Experience> episode )
    int t=0;
    while (!episode.isEmpty()){
        Experience e = episode.removeFirst();
        String s_e = e.getS();
        int a_e = e.getA();
// G = r_k+gamma^1*r_{k+1}+gamma^2*r_{k+2}+... until end of episode
        double G=e.getR(); // gamma to the power of 0 is 1
        ListIterator iterator = (ListIterator)episode.iterator();
        int k=t+1;
        while(iterator.hasNext()){
            Experience fe = (Experience)iterator.next();
            G+=Math.pow(GAMMA,k-t)*fe.getR();
            k++;
        }double
        [] pi_sa = P_Policy(s_e);
        double[] theta = getTheta(s_e);
        List <Integer> A_s = env.coursesOfAction(s_e);
        double gradient_ai = 0;
        for (int a_i : A_s){
            gradient_ai=-pi_sa[a_i];
            if (a_i==a_e) gradient_ai=gradient_ai+1;
                            theta[a_i]   +=   ETA_theta*Math.
pow(GAMMA,t)*G*gradient_ai;
        }
        setTheta(s_e,theta);
        t++;
    }}
```

In the Monte Carlo algorithm in ► section "Subsequent evaluation of episodes ("Monte Carlo" Method)", we computed the state scores using a goal-based approach that determines the respective state scores starting from the end of the episode. In tactic search, we are concerned with the prediction of rewards. Therefore, in the first case, we computed "from the end," while this time we computed from the beginning, cumulating the incoming discounted rewards in each case.

REINFORCE shows a clear learning progress. The step size parameter η_θ must be chosen appropriately. However, REINFORCE can have a high variance and may become slow as a result. To reduce the variance, we need to better "orient" the agent's learning toward the goal.

Monte Carlo Policy Gradient with State-Value Function as a Guide (REINFORCE with Baseline)

We orient the learning process with the help of a so-called baseline, which provides a kind of orientation value for the targeted state evaluation.

$$\theta_{t+1} = \theta_t + \eta_\theta \left[G_t - b(s_t) \right] \nabla \ln \pi (s,a,\theta) \qquad (4.17)$$

With such a baseline, the policy is adjusted particularly strongly if there is a large deviation between the orientation value and the reward actually collected. Because this baseline can also be uniformly zero, REINFORCE can be seen as a special case of this update.

For our purposes, it makes sense if this orientation value is variable. For example, in some states actions have very high valuations, and there a too small orientation value would have only a small effect. In other states, on the other hand, a lower base value is appropriate.

It is therefore natural to use the current estimate of the state value for this "baseline." $\hat{V}(s_t)$ to be used for this "baseline":

$$\theta_{t+1} = \theta_t + \eta_\theta \left[G_t - \hat{V}(s_t) \right] \nabla \ln \pi (s,a,\theta) \qquad (4.18)$$

Thus, the policy is adjusted particularly strongly when the discounted and accumulated reward G observed in the aftermath, i.e., until the end of the episode, shows a large deviation from the currently given state value estimate. We saw in Chap. 3 how tactic improvement and valuation optimization can optimize each other. If the value estimate and observed outcome of the exploration episode match, then little adaptation takes place. This stabilizes the learning process.

Monte Carlo Policy Gradient Algorithm REINFORCE with baseline

```
1      π is a differentiable stochastic policy parameterization π(s,a,θ)

2      V̂(sₜ) an approximation of the state-value function

3      ηθ > 0, ηᵥ > 0 learning parameters (step size)

4      Initialize policy parameter θ ∈ ℝᵈ and state-value function (e.g. 0)

5      Loop for each episode:

6          Generate an episode πθ: s₀,a₀,r₁, s₁,a₁,r₂, ... ,s_{T-1},a_{T-1},r_T with πθ

7          Loop for each step of the episode t = 0,1,...,T − 1:

8              G ← Σᵏₖ₌ₜ₊₁ γᵏ⁻ᵗ⁻¹ rₖ

9              δ ← G − V̂(sₜ)

10             V̂(sₜ) ← V̂(sₜ) + ηᵥδ  (tabular state-value function)

11             θ ← θ + ηθγᵗ δ∇ ln π(sₜ,aₜ,θ)
```

The following is an implementation of the algorithm in Java.

Java implementation of an (episodic) 'REINFORCE with baseline' update (tabular policy parameterization and state-value approximation)

```
protected void update ( LinkedList <Experience> episode )
 int t=0;
 while (!episode.isEmpty()){
    Experience e = episode.removeFirst();
    String s_e = e.getS();
    int a_e = e.getA();
// G = r_k+gamma^1*r_{k+1}+gamma^2*r_{k+2}+... until end of episode
    double G=e.getR(); // gamma to the power of 0 is 1
    ListIterator iterator = (ListIterator)episode.iterator();
    int k=t+1;
    while(iterator.hasNext()){
        Experience fe = (Experience)iterator.next();
        G+=Math.pow(GAMMA,k-t)*fe.getR();
        k++;
    }
    double v = getV(s_e);
    double advantage = G-v;
    double v_new = v + ETA_V*advantage;
    setV(s_e, v_new);
    double[] pi_sa = P_Policy(s_e);
    double[] theta = getTheta(s_e);
    List <Integer> A_s = env.coursesOfAction(s_e);
    double gradient_ai = 0;
    for (int a_i : A_s){
        gradient_ai=-pi_sa[a_i];
        if (a_i==a_e) gradient_ai=gradient_ai+1;
        theta[a_i] +=
        ETA_theta*Math.pow(GAMMA,t)*advantage*gradient_ai;
    }
    setTheta(s_e,theta);
    t++; }
}
```

It is not trivial to find the optimal values for the step size. Although high values accelerate the learning process, a too high step size can lead to the fact that no optimal performance is found or even unfavorable "dead ends" or "circles" get stuck. By introducing the baseline, the learning process can be accelerated significantly. ◉ Figure 4.10 shows test runs of the presented implementation in the arena "mapWith-Trap3" with intuitively selected parameters.

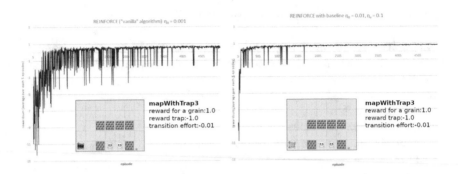

Fig. 4.10 Example runs of the implementation of Monte Carlo policy gradient REINFORCE with and without baseline in the arena "mapWithTrap3"

4.2.3 Combined Methods (Actor-Critic)

"Actor-Critic" Policy Gradients

The difference $\left[G_t - \hat{V}(s_t) \right]$ reminds us of the temporal difference, as we note the difference between observation and estimation. However, so far we have followed the Monte Carlo approach by evaluating full episodes retrospectively. Can we also use the policy gradient to perform learning "online" and process observations directly?

At the level of individual time steps, we only have the instantaneous TD error available. We therefore modify Gl. 4.18 so that we use the TD error for the magnitude of the online adjustment:

$$\theta_{t+1} = \theta_t + \eta_\theta \left[r_{t+1} + \gamma \hat{V}(s_{t+1}) - \hat{V}(s_t) \right] \nabla \ln \pi (s, a, \theta)$$

In this way, we combine the policy-gradient tactic search with a goal-oriented evaluation optimization by processing the temporal difference in different ways online. The temporal difference is now processed "simultaneously" in two places. First, in the direct policy improvement, and second, in the estimation improvement $\hat{V}(s)$. The corresponding algorithms are the so-called "Actor-Critic" algorithms.

"Actor-Critic" (Tabular Policy and State-Value Approximation))

```
1        π_θ a stochastic policy parameterization π_θ(s)
2        V̂(s_t) an approximation of the state-value function
3        η_θ > 0, η_v > 0 learning parameters (step size)
4        Initialize policy parameter θ ∈ ℝ^d and state-value function (e.g.to 0)
5        Loop for each episode:
6              Initialize s with start state
7              I ← 1
8              Loop while s is not terminal
9                    choose a ~ π_θ(s)
10                   get s' and r from the environment (observation)
11                   δ ← r + γV̂(s') − V̂(s)
12                   V̂(s) ← V̂(s) + η_v δ (critic with tabular implementation of V̂)
13                   θ ← θ + η_θ I δ∇ ln π_θ(s)
14                   I ← γI
15                   s' ← s
```

The "actor-critic" conception basically provides for two separate components: one that tactics-based controls the agent's behavior (actor) and another that evaluates the behavior (critic).

The "actor-critic" algorithms thus learn both a policy π as well as a value function V "online." The "actor" is a component that learns the policy, while the "critic" is a component that learns about the policy.

The "critic" uses a standard TD algorithm to learn a state score. In principle, the TD error can be used to assess whether a chosen action was "good" or "bad" in terms of the average results achieved so far. An action was "well" chosen if it led to a state with a better than expected value and "bad" if it led to a state with a lower than expected value. With this feedback from the "critic" based on the TD error, the "actor" adjusts its control using the policy gradient, which increases the probability of favorable actions and decreases that of unfavorable actions. At the same time, the critic also uses the TD value to improve its state evaluation.

A key feature is the division of labor with respect to reactive execution and independent evaluation of outcomes: the "actor" does not have direct access to the reward signal, and the "critic" does not have direct access to the action selection.

Java Implementation of an "actor-critic"-Update (Tabular State-Value Function and Policy)

```java
protected void update(String s, int a, double reward, String s_
new, boolean
episodeEnd ){
  double observation = 0.0;
```

```
if (episodeEnd) {
    observation = reward;
} else {
    observation = reward + (GAMMA * getV(s_new));
}
double v = getV(s);
double delta = observation-v; // temporal difference
// update "critic
double v_new = v + ETA_V*delta;
setV(s, v_new);
// update "actor
double[] pi_sa = P_Policy(s);
double[] theta = getTheta(s);
List <Integer> A_s = env.coursesOfAction(s);
double gradient_ai = 0;
for (int a_i : A_s){
    gradient_ai=-pi_sa[a_i];
    if (a_i==a) gradient_ai=gradient_ai+1;
    theta[a_i] += ETA_theta*I_gamma*delta*gradient_ai;
}
setTheta(s,theta);
I_gamma  = GAMMA*I_gamma;  // update discount factor (global
variable)
}
```

Technical Improvements to the Actor-Critic Architecture

Research in the area of reinforcement learning is currently intense, almost frantic. Recently, a large number of further developments have been presented that improve the learning process by better exploiting statistical methods and technical properties of the hardware. One of the newer algorithms will be presented in the following as an example. It will also become clear how improvements to the described basic architecture can be achieved.

For both the "REINFORCE with baseline" and "actor critic" algorithms, we have changed our definition of performance. We have chosen an adaptation function that focuses on minimizing the "surprises" that occur when exploring the environment. Instead of maximizing the absolute reward, or score, of the starting state, we seek to minimize the observed error by adjusting π and $\hat{V}$ as the exploration progresses.

$$VE_{p_q,v}(s) = (R - V(s))^2$$

The two algorithms mentioned above can technically be seen as two opposite extreme cases. On the one hand, in "REINFORCE with baseline" R is formed by looking to the end of the episode to see what the cumulative and discounted reward with $R_t = G_t = \sum_{i=1}^{T_{max}} \gamma^{i-1} r_{t+i}$ will be in total. In the other case with "actor-critic," we only take a single step, in which case the value $R_t = r_{t+1} + \gamma \hat{V}(s_{t+1})$ is formed according to the Bellman equation from the collected reward and the discounted valuation of the successor state.

In practice, a middle ground between the two extremes has proven useful. Instead of exploring Monte Carlo style to the end of the episode or, in the other case, looking only a single step further, we can look forward a certain number of n steps and use a so-called n-step return (Fig. 4.11):

$$R = \sum_{i=0}^{n-1} \gamma^i r_{t+i} + \gamma^n \hat{V}(s_{t+n}) \tag{4.19}$$

It's also referred to as the "advantage function." $A_{\pi_\theta, V}(s_t, a_t)$ is used:

$$A_{P_q, V, n}(s_t, a_t) = \sum_{i=0}^{n-1} g^i r_{t+i} + g^n \hat{V}(s_{t+n}) - \hat{V}(s_t) \tag{4.20}$$

In principle, the observations are accumulated a few steps in advance and set in difference to the existing estimate. In the one-step case, the advantage function corresponds exactly to the known TD error:

$$A_{P_q, V, 1}(s_t, a_t) = r_{t+1} + g \hat{V}(s_{t+1}) - \hat{V}(s_t) \tag{4.21}$$

Furthermore, it is technically convenient if several agents working simultaneously could explore the environment in parallel and gather their experiences in a shared structure.

One algorithm that works on these challenges is the asynchronous advantage actor-critic (A3C) (Mnih et al. 2016), also abbreviated as "A3C." A tabular version of the algorithm is presented below. Moreover, the parallel collection of

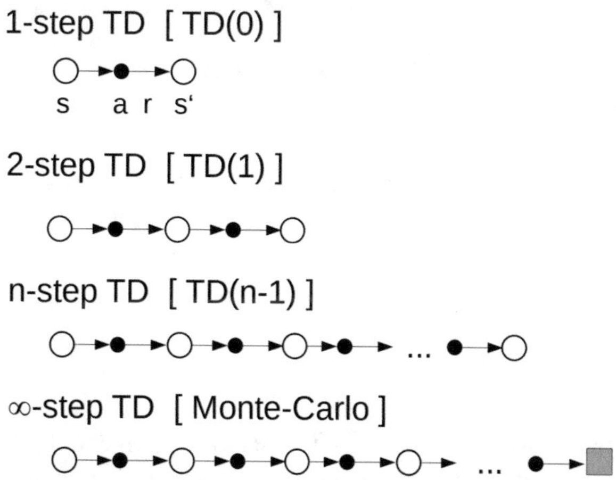

Fig. 4.11 "Backup diagrams" showing the connection of n-step TD algorithms with the Monte Carlo update

independent observations has a particularly beneficial effect when using approximators such as artificial neural networks (cf. Chap. 5).

The algorithm allows the required computing time to scale with the number of threads working in parallel.

The parallel exploration of the state space sometimes also brings favorable effects, which are not only due to the better utilization and combination of the available computing power, but can be explained by an improved exploration process. However, the great advantage of these algorithms only becomes clear when using approximators such as artificial neural networks (Chap. 5).

Asynchronous Advantage Actor-Critic – Pseudocode for Each Actor-Learner Thread (Mnih et al. 2016).

1 Assume global shared policy parameter vectors θ and a global state-value approximation $\hat{V}$ and a global shared step counter T= 0

2 Assume thread-specific parameter vectors θ' state-value approximation $\hat{V}'$

3 Initialize thread step counter $t \leftarrow 1$

4 repeat

5   Reset gradients: $d\theta \leftarrow 0$ and $dv \leftarrow 0$

6   Synchronize thread-specific parameters $\theta' = \theta$ and $\hat{V}' = \hat{V}$

7        $t_{start} = t$

8   Get state s_t

9   repeat

10   choose $a_t \sim \pi_\theta(s_t)$

11   Receive reward r_t and new state s_{t+1}

12               $t \leftarrow t + 1$

13               $T \leftarrow T + 1$

14   until terminal s_t or $t - t_{start} == t_{max}$ // terminal or n steps solved?

15                               $R = \begin{cases} 0 & for\ terminal\ s_t \\ \hat{V}' & for\ non-terminal\ s_t \end{cases}$

16   For $i \in \{t-1, ..., t_{start}\}$ do

17               $R \leftarrow r_i + \gamma R$

18   Accumulate changes $\Delta\theta'_{s_i} \leftarrow \Delta\theta'_{s_i} + \eta_\theta (R - \hat{V}'(s_i))\nabla_{\theta'} \ln \pi(a_i, s_i, \theta')$

19   Accumulate changes $\Delta\hat{V}_{s_i} \leftarrow \Delta\hat{V}_{s_i} + \eta_v (R - \hat{V}'(s_i))$ (tabular state values)

20   end for

21   Perform asynchronous update of θ and $\hat{V}$ using $\Delta\theta'_s$ and $\Delta\hat{V}_s$ for all changes stored.

22 until $T > T_{max}$

A first change is that instead of an agent, we have a "population," but some of the data structures are shared. These shared data structures have been implemented here as static class attributes in A3C_Hamster.

```
protected static HashMap <String, double[]> thetas_global;
protected static HashMap <String, Double> V_global;
protected static int cnt_steps_global=0;
protected static int cnt_episodes_global=0;
protected static int max_episodes_global=250000;
protected static double sum_reward_global=0.0;


public static int NUM_A3C_AGENTS = 8;
public      static      A3C_Hamster[]      hamsters      =      new
A3C_Hamster[NUM_A3C_AGENTS];
```

Java Implementation of an "A3C"-Update (Tabular State-Value Function and Policy)

```
boolean episodeEnd ){
 if (nstep_sequence.isEmpty()) return;
    LinkedList     <StateThetas>     delta_theta=new     LinkedList
<StateThetas> ();
 LinkedList <StateV> delta_V = new LinkedList <StateV> ();
 double R = 0;
 if (!episodeEnd){
    R=getV(s);
 }
 while (!nstep_sequence.isEmpty()){
    e = nstep_sequence.removeLast();
    String s_i = e.getS();
    int a_i = e.getA();
    R=e.getR()+GAMMA*R;
    double advantage = R-getV(s_i);
    // accumulate gradients
    double[] pi_sa = P_Policy(s_i);
    double[] d_theta_s = new double[SIZE_OF_ACTIONSPACE];
    double d_v=0.0;
    ArrayList <Integer> A_s = env.coursesOfAction(s_i);
    for (int b : A_s){
        double gradient_b=-pi_sa[b];
        if (b==a_i) gradient_b=gradient_b+1;
        d_theta_s[b] = ETA_theta*gradient_b*advantage;
        d_v = ETA_V*advantage;
    }
    delta_theta.add(new StateThetas(s_i,d_theta_s));
    delta_V.add(new StateV(s_i,d_v));
 }
}
```

The "work" is done in parallel working threads implemented as a static array at the class A3C_Hamster. The run method is executed in the worker thread. The asynchronous update of the data structures takes place in the act method.

```
@Override
public void run() {
  int cnt_done=0;
done=true; while (!Thread.currentThread().isInterrupted()){
    if (!done) {
        act_task();
        cnt_done++;
            if ((cnt_done>ACTS_BEFORE_UPDATE)||(episodeEnd)){  //
default 0
            done=true;
            cnt_done=0;
        }
    }
} System.out.println(workers[ID].getName()+": Bye, bye.");
}
@Overridepublic
void act(){
 if (env==null) return;
if (done) {
    // update global parameters
    updateGlobalTheta();
    updateGlobalV();
    if (episodeEnd) startNewEpisode();
    done = false;
}
}
```

Numerous other improvements to the presented architecture have been proposed, which take better account of the technical properties of the computational technology and the stochastic properties of the accruing data sets. A comprehensive collection of the current algorithms with policy gradients and the corresponding papers can be found, for example, at https://lilianweng.github.io/lil-log/2018/04/08/policy-gradient-algorithms.html (31.5.2021).

In the "DynaMaze" card with the parameters

reward for a grain = 1.0 (terminal)
reward trap = -1.0 (terminal)
transition effort = -0.01learning
paramter: $\eta_v = 0.0$ $\eta_\theta = 0.005$ $\gamma = 0.9999999$ $T = 1.0$

on an AMD Ryzen 7 2700X eight-core processor in the Greenfoot Standard VM, test runs with 250,000 episodes resulted in runtimes of about 7 seconds for the Actor Critic and about 3 seconds for the A3C. The A3C ran with eight threads and seven-step advantage function. The average accumulated reward was 0.85 per episode in both cases. With 1,000,000 episodes, this resulted in about 25 seconds for the simple AC and 10 seconds for the A3C.

The runtimes do not scale linearly with the number of threads due to the increased overhead and the properties of our environment. The profiling tool visualvm offers a possibility to further tune the resource consumption. It can be used to examine in detail, for example, the dwell times in the respective functions, but also memory consumption and much more. The tool is available free of charge at (https://visualvm.github.io/index.html; 2.6.2021). Certainly, further improvements can be realized, e.g., by allowing the threads to execute several episodes independently of act and by adjusting their rhythm to the simulation speed of Greenfoot. This kind of programming is not really foreseen in the Greenfoot environment, which is designed for educational purposes, so it may cause problems, e.g., with mouse handling. The technical tricks with which one or the other delta can still be extracted should not interest us further at this point for the time being. A total of 100,000 episodes per second are, I think, not bad for such a didactic environment. The aim of the book is to provide an overview of the various basic approaches to reinforcement learning, and at this speed one or two questions can certainly be investigated quite well.

Feature Vectors and Partially Observable Environments

So far we have assumed Markov systems where we have full access to the states. For ordinary beings with limited observables, decisions may involve "surprises," since in each case the same action in superficially identical states (observables) may produce entirely different results (cf. ● Fig. 4.12).

An embedded agent which, as shown in ● Fig. 4.12 has only limited observation possibilities, cannot recognize from its immediate perception whether it can find a reward when it moves around the front-left corner.

One solution would be to preface a state estimator with "latent states," which helps to better assess the situation, e.g., using action and observation history to

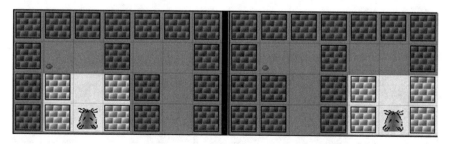

Fig. 4.12 For an agent with limited observation possibilities who assumes that the world consists only of observable states, it remains inexplicable why he finds a grain in state s with the action sequence ("north," "north," "west") once and not once

determine: I am at location A or at location B. This requires increased demands on the modeling capabilities of our agent. Here we also leave the framework of Markov decision processes (MDPs) and arrive at the setting of partially observable Markov decision processes (POMDP). There is still a lot of research going on in this area (Sutton and Barto 2018, Chap. 17.3) The problem of state estimation, occurs especially in the field of robotics. In RoboCup, for example, the robots playing soccer have to "guess" as correctly as possible where they are on the field. In the early phase of RoboCup, this problem also led to many "own goals," as the robotic players shot at the wrong goal.

With policy gradients, we can generate explicit stochastic policies that we continuously adapt. Therefore, the algorithms with stochastic ("SoftMax") policies have advantages in uncertain or only partially observable environments. Stochastic policies allow us to use model-free reinforcement learning to increase the reward even under uncertain conditions. Although this approach does not achieve the value that would theoretically be possible with an environment model, we can significantly increase the performance because the stochastic policy can assign precisely adjusted probabilities to the actions, which correspond better to the properties of the environmental system.

This is particularly useful in "uncertain" scenarios where it does not make sense to assign a single "optimal" action decision to a given state. For example, consider the bacterium *Escherichia coli* from ▶ section "Monte Carlo tactics search". From our perspective, it is in a comparatively huge environment (e.g., an intestine). However, the bacterium's world consists of only a very small state space. This forces it to make "stochastic" decisions. The situation is similar in a poker game, although we can make sense of the hands we pick up, we cannot gain insight into the psychology of the players. Which is why in certain situations we can only choose decisions according to a certain probability distribution. The example in ◉ Fig. 4.13,

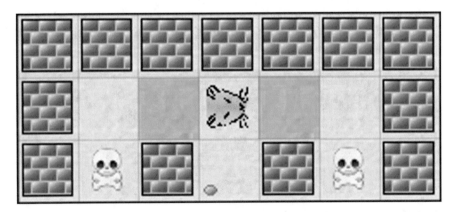

Fig. 4.13 The hamster is in an uncertain environment. The blue tinted fields are indistinguishable for the agent if it only has the environment information (adjacent wall present or not)

borrowed from David Silver (https://www.youtube.com/watch?v=KHZVXao4qXs, 20.3.2020), illustrates the problem.

Let us further assume that the agent's states and their evaluation emerge only from certain "selected" features of its observable environment. To realize this, we introduce a feature vector x which tells us the presence or absence of certain properties in the agent's environment, such as the presence of a wall. The state space is thus in principle formed by the states of the feature vector.

In our example case, we can only see if there is a wall to the north, east, south, or west of the hamster; the directions do not rotate with it in our example. It is therefore a kind of spotlight that only illuminates the parts of the world that are directly adjacent to the agent:

$$x_i = \begin{cases} 1, & \text{if there is a wall} \\ 0, & sonst \end{cases} \qquad i \in \left\{ "Nord"(0), "East"(1), "South"(2), "West"(3) \right\}$$

For both marked fields s, we thus obtain the vector:

$$x(s) = (1,0,1,0)^T$$

The policy parameters θ_a are now to form the preferences for each action by the linear product of the parameter vector and the feature vector. $h(s,a,\theta_a) = \theta_a^T x(s)$ for each action. Each action would thus have its own parameter set with which the assignment is calculated.

However, we can also combine this directly with the possible actions to be able to parameterize action preferences for complete state (or feature) action pairs in each case. To do this, we add an "action vector" to the feature vector, which represents the available motor activity. We build a mapping of the possible actions for our hamster agent in the vector such that we have:

$$x_j = \begin{cases} 1, & \text{if } j = a \\ 0, & otherwise \end{cases} \qquad j \in \left\{ "North"(0), "East"(1), "South"(2), "West"(3) \right\}$$

set. For the action $a = "Ost"$ we got the vector: $x(1) = (0,1,0,0)^T$ This would have put the action $a = "Ost"$ in one of the light marked fields s would have the representation:

$$x(s, "East") = \underbrace{1,0,1,0}_{\substack{sensors \\ (walls)}}, \underbrace{0,1,0,0}_{\substack{motors \\ (direction)}})^T$$

Following the formulation in (Sutton and Barto 2018, Chapter 13), we can now also use the term that is only slightly more complicated compared to Gl. 4.16 term, which is only slightly more complicated:

$$\nabla \ln \pi(s,a,\theta) = x(s,a) - \sum_b \pi(s,b,\theta)x(s,b) \qquad (4.22)$$

for the policy gradient. It results mathematically in a similar way as Gl. 4.16. Although we are now using a parameter number that is probably much too large, we can still use this later in the section where we talk about the use of approximators.

Calculating the Actor-Critic Update from Feature Vectors in Java

```
protected void update( String xs_key, int a, double reward, String
xs_new_key, boolean episodeEnd ){
 double observation = 0.0;
if (episodeEnd) {
    observation = reward; } else {
    observation = reward + (GAMMA * getV(xs_new_key)); } double v =
getV(xs_key); double delta = observation-v;
 // Update "critic" double v_new = v + ETA_V*delta; setV(xs_key,
v_new);
 // Update "actor" double[] x_sa = getFeatureVector(xs_key, a);
double[]
theta_xa    =    getTheta(xs_key,a);    double[]    gradient=
gradient_ln_pi(xs_key,a);
for (int k=0;k<theta_xa.length;k++){
    theta_xa[k] += ETA_theta*I_gamma*delta*gradient[k];
} setTheta(xs_key,a,theta_xa);
I_gamma = GAMMA*I_gamma;
}
private double[] gradient_ln_pi(String xs, int a){
 double[] gradient = this.getFeatureVector(xs,a);
double[] pi_s = P_Policy(xs); for (int k=0;k<gradient.length;k++) {
    double sum=0;
    for (int b=0;b<pi_s.length;b++){
        double[] x_sb = this.getFeatureVector(xs,b);
        sum+=pi_s[b]*x_sb[k];
    }
    gradient[k]-=sum;
 }
return gradient;
}
```

Using the solutions found by the algorithms for the example with ambiguous states ◉ (Fig. 4.13), the difference between an evaluation-based and a policy-based learning can be well illustrated (cf. ◉ Fig. 4.14).

The stochastic policy generated by the (SoftMax) policy gradient therefore works, significantly better than a "winner takes all" policy, such as the one generated by the Q-learning ◉ (Fig. 4.15).

While with the policy gradient on the indistinguishable fields for the actions "west" and "east," respectively, the probabilities are equal with values 0.5.

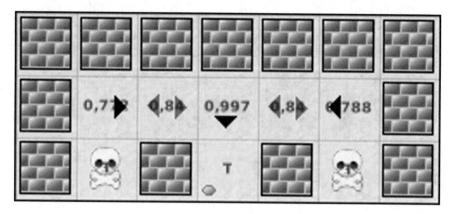

Fig. 4.14 Result of the "actor-critic" with the described feature vectors. On the ambiguous fields, the stochastic policy sets equal probabilities for the action "west" and "east" with 0.5 each

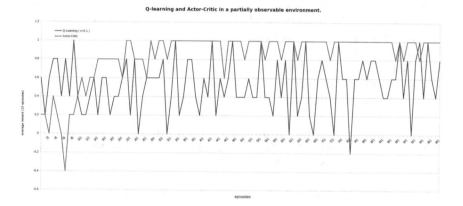

Fig. 4.15 Comparison of the average obtained reward in the Java example of "Q-learning" and "actor-critic" policy gradient (in 10 test runs each)

Q-learning recognizes equivalent valuations, but the agent must decide on the same "optimal" direction in the two indistinguishable fields, which the agent follows with probability $p^*(a|s) = 1 - \varepsilon$ follows. Therefore, the success rate falls well short of the stochastic tactic generated with the policy gradient ⦿ (Fig. 4.16).

Summary
With Q-learning and SARSA learning, we have learned about value-based learning methods. These learn an evaluation function to determine the best decision for a given state in each case. The control itself is not part of the learning and is designed to be more or less "greedy" to ensure exploration.

In the tactics search procedures, the policy is learned. The evaluation function plays at most a supporting role. Parameterizations of the policy allow a "soft"

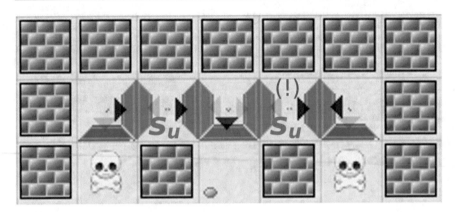

Fig. 4.16 For the uncertain states s_u, evaluation-based learning with implicit ε-greedy policy develops a suboptimal solution

adaptation of the control. Thus, stochastic policies can be generated and optimal probability distributions can be searched for in each case. This is particularly interesting in partially observable worlds, as it allows us to account for the properties of the non-observable part of the world in our policy to some extent.

In "actor-critic," both the evaluation function and the policy are learned. Both components support each other, with the resulting behavior being generated by a policy that is adapted according to a policy gradient procedure.

4.3   Exploration with Predictive Simulations ("Model-Based Reinforcement Learning")

All algorithms presented so far belonged to the so-called "model-free" methods. They neither use preset knowledge about the environment nor do they create an explicit model of their environment. The algorithms generated controls that derive their favored actions from a given state via stored evaluations or from direct assignments, respectively.

Learning thus takes place through a constant re-evaluation of past states. Observations are processed in such a way that the difference between the "reward" expected in the past and that actually received is reduced. In this way, the algorithms generate a kind of "reactive competence." The model-free methods learn "retroactively," so to speak. The agent thus always "knows" immediately which action is to be chosen.

However, one can quickly convince oneself that models that allow quasi "inner simulations" can be immensely useful. Good models allow an agent to predict how its actions would affect the environment. This makes it possible to use "thought experiments," so to speak, to make purposeful predictions without actually having to act. We can thus produce "simulated experiences" with models.

Real experience has a much greater value, since it is ultimately what matters, but the benefit of planning by means of a kind of "map" quickly becomes obvious when one considers the costs that real experimentation can entail. To put it in the manner of Karl Popper: hypothetical behavior allows that instead of the real agent only its hypothesis perishes.

A current topic of AI research is to unify the two approaches of "retroactive adaptation" of memory and "predictive planning" using internal models (Sutton and Barto 2018, Ch. 8).

Since "GOFAI" also deals with searching and planning by means of knowledge about the outside world ("knowledge-based systems"), there are extensive works on this, such as Russell and Norvig (2010). Unfortunately, we can only examine a few samples, with the main aim of closely linking "improving response," on the one hand, and "planning ahead," on the other. In the following example, we will see that much more information can be extracted from single observations if we remember them and use them to perform numerous low-cost "virtual actions."

4.3.1 Dyna-Q

The idea is to create virtual experiences by repeating individual actions of the past in our respective provisional "model." In each case, the model provides us with the subsequent state seen in the past and any direct rewards that may be present. With these "thought experiments" we improve our state-action evaluations "on a large scale" without having to make the corresponding observations in reality.

If a state is to be predicted with a model s' the initial situation in the model can be chosen arbitrarily. (s, a) in the model can be chosen arbitrarily. Predictions can therefore be produced decoupled from the current state or action evaluations and the current "action policy" of the agent. We can also explore the environmental system within the framework of our model independently of real experiences and thus perhaps discover better paths or evaluations "virtually" at first.

The DynaQ algorithm (Sutton and Barto 2018) is very exciting because it represents an approach, processes information gained from interaction with the environment at each step "online," and integrates "model learning" and model-free reinforcement learning. The "online experiences" influence the model and thus also permanently change the predictive planning.

Within an agent with planning there are at least two uses for "real" experiences that come from the environment: The first is that they can be used directly to improve the evaluation function or policy, and the second is that they can be used to improve the model. In the former use, we can use direct reinforcement learning, as described in the previous sections. In the second use, the aim is to better fit the model to the real environment

The relationships between the components of the learning system are shown in Fig. 4.17 depicted. The arrows mark effects that influence and, if possible, improve the corresponding components. This has inspired many further developments, for example, some of the outstanding works of David Silver.

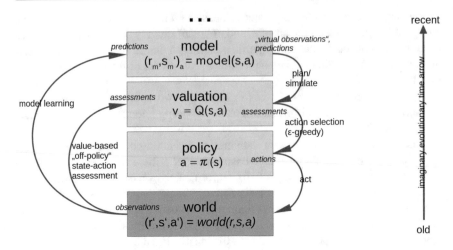

Fig. 4.17 Relationships of the components of the learning system at Dyna-Q

"Real-world" experiences can improve the state assessment and policy either directly or also indirectly through the model. The latter is sometimes referred to as "indirect reinforcement learning." Indirect methods make better use of the limited amount of experience and thus achieve better control with less effort. On the other hand, direct methods are much simpler and are not affected by biases and by the design of the model.

A distinctive feature of Dyna-Q is that it combines two fields of AI research, one being "deliberative planning," more of a "GOFAI" approach, and the other being "reactive decision making," an approach that comes from machine learning and emphasizes the "situatedness" and "physicality" of agents.

The algorithm presented in (Sutton and Barto 2018) is minimalistic, which suits the goal of this book to make the learning algorithms understandable. The planning takes place as a single-stage, tabular Q-scheduling method at random. Single-stage tabular Q-learning is used for the direct reinforcement learning. The model learning method is also table-based and assumes that the environment behaves deterministically. After each transition $s_t, a_t \rightarrow r_{t+1}, s_{t+1}$, a table entry is created that allows to predict which state and which reward would follow an eventual action a in state s. Thus, when the model is queried with a previously stored state-action pair, it simply returns the last observed subsequent state and its associated direct reward as its prediction. When "planning," the algorithm randomly accesses previously observed state-action pairs. For learning from real experiences as well as for indirect learning from simulated experiences, Dyna-Q uses the same reinforcement learning algorithm. "Learning" and "planning" are deeply integrated in Dyna-Q. They really only differ in the source of their experience.

In principle, planning, acting, modeling, and direct reinforcement learning in dyna-agents can take place in parallel. However, for execution on a serial computer, these can also be executed sequentially within a time step. In Dyna-Q, the processes of acting, model learning, and direct RL require relatively little computational effort. Most of the time is spent in the planning process. The algorithm is designed to perform a certain number of iterations of the Q-planning algorithm in each step after acting, model learning, and direct reinforcement learning.

In the pseudocode algorithm for Dyna-Q, model(s, a) denotes the content of (predicted next state and reward) for the state-action pair (s, a). Direct reinforcement learning, model learning, and planning are implemented by steps (d), (e), and (f). If (e) and (f) were omitted, the remaining algorithm would be a one-step tabular Q-learning.

"Dyna-Q" (Sutton and Barto)

```
1 Initialize Q(s,a) and Model(s,a)
2 Loop for each episode:
3      Initialize start s
4      Repeat
5           s ← current state
6           a ← πQ(s) (e.g. ε-greedy)
7           Take action a and observe r and s'
8           Q(s,a) ← Q(s,a) + η[r + γ max Q(s',a') − Q(s,a)]
                                        a'
9           Model(s,a) ← r,s' (deterministic environment)
10          Loop repeat n times:
11               s ← random previously observed state
12               a ← random action previously taken in S
13               r,s' ← Model(s,a)
14               Q(s,a) ← Q(s,a) + η[r + γ max Q(s',a') − Q(s,a)]
                                           a'
```

The scenario provided in the folder "Chapter 4 Decision-Making and Learning\ HamsterWithTDLearning" also contains an agent "DynaQHamster." In addition to this hamster agent, if you also set the arena dynaMaze by commenting in the constructor of the class "TD_AgentEnv" the lines "hamster = new DynaQHamster();" and "super(dynaMaze);," you can reproduce the "Dyna Maze" task from Sutton and Barto (2018) in our hamster model. The implementation of Dyna-Q complements Q-learning with the "model learning" described in the pseudocode starting at line 8.

Dyna-Q in Java

```
@Override
public void act(){
 if (s_new==null) {
    s = getState(); }else{
    s = s_new; }
 // apply policy double[] P = P_Policy(s); a =
selectAccordingToDistribution(P); incN(s,a);
```

```
    // apply transition model (consider uncertainties in the
result of an
// action)
 int dir = transitUncertainty(a);
 // execute action a if (dir<4){
    try{
        setDirectionOfView(dir);
        goAhead();
    }catch (Exception e){
        //System.out.println("bump!");
    }
    cnt_steps++; } // get new state s_new = getState();
  // get the reward from the environment double r = env.
getReward(s_new);
sum_reward+=r; episodeEnd = false; if
((env.isTerminal(s_new))||(cnt_steps>=max_steps)) {
    episodeEnd=true; }
 // Q-update update(s,a,r,s_new, episodeEnd);
 // Update model setToModel(s,a,new Observation(s_new,r));
 // Generate simulated experiences "planning" for (int
i=0;i<this.planningIterations;i++){
    simulateAnExperience(); }
  if (this.evaluationPhase) env.putTracemarker(getSX(s),getSY(s
),a,1.0); if
((env.DISPLAY_UPDATE)&&(cnt_steps%env.
DISPLAY_UPDATE_INTERVAL==0))
env.updateDisplay(this);
 // episode end reached? if (episodeEnd) {
    startNewEpisode(); }
}
public void simulateAnExperience(){ String[] s_keys =
model.keySet().toArray(new String[0]); String s_sim =
s_keys[random.nextInt(s_keys.length)];              HashMap
<Integer,Observation>
predictions = model.get(s_sim); Integer[] actions =
(Integer[])predictions.keySet().toArray(new
                Integer[0]); int a_sim =
actions[random.nextInt(actions.length)]; Observation o =
predictions.get(a_sim); update(s_sim,a_sim,o.getR(),o.getS(),
                env.isTerminal(getSX(o.getS()),getSY(o.getS()))));
}
```

By "planning" with the model, the observations of the past can become useful again. This is also practically shown by the learning curves depending on the planning steps performed ◉ (Fig. 4.18). The principle pays off especially where "real"

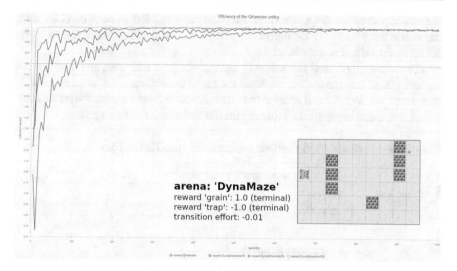

Fig. 4.18 Learning curves of the hamster agent with DynaQ in "DynaMaze" as a function of the number of planning steps performed. ($\varepsilon = 0.1$; $\eta = 0.01$; $\gamma = 0.9999$)

experiences are relatively "expensive," e.g., in robotics, or where the largest possible areas of the state space are to be evaluated with comparatively few real observations (e.g., in certain board games).

Dyna-Q can thus make better use of the agent's online experiences through the "virtual" repetitions in the model than would be the case, for example, with simple Q-learning. It can be seen in the hamster scenario that Dyna-Q performs significantly better in comparison with the algorithms that form evaluation functions without a model.

4.3.2   Monte Carlo Rollout

In the following example, we will use the model to perform not only a certain number of "virtual actions" but complete "virtual episodes." The so-called "rollout algorithms" evaluate the respective action options by determining the average returns from numerous simulation runs. It is helpful that we can generate these arbitrary episodes at very low cost and therefore in very large numbers. For this, we need an inner model that allows us to execute "virtual moves" and predict possible states. The space of possible states and paths quickly grows immeasurably without heuristic tools. It is practically very seldom possible to search this space "blindly" schematically, as described in Chap. 3.

In rollout algorithms, simulations are started for each possible action in the given state and then continued with random progressions or a "cheap" rollout policy. Unlike "real" Monte Carlo algorithms, rollout algorithms are not concerned with producing a complete evaluation function $\hat{Q}(s,a)$. The Monte Carlo estimates of the action values are determined only for the state at hand in each case. In many cases, it is not possible or even necessary to produce and store state or action scores for all states visited or simulated.

The rollout algorithms immediately use the obtained estimates and immediately discard them. This makes the implementation of the rollout algorithms much simpler, since we do not need to approximate a function over the entire state or state-action space and compute adjustments for the stored state-action pairs.

Pseudocode "Monte Carlo rollout evaluation" (for TicTacToe)

```
1 function rollout_evaluation(s,player)  ;s ∈ S and player ∈ {'X','O'}

2      if s a terminal state

3             return Reward(s)

4      else

5             opponent ← {'X','O'} \ player

6      perform random action a and observe s' (simulation).

7             v ← −rollout_evaluation(s',opponent)

8             return v
```

A Java implementation of the algorithm in Greenfoot can be found in the companion material in the directory "Chapter 4 Decision-Making and Learning\ TicTacToeWith_Q_MCRollout_MCTS."

The recursive rollout has great structural similarities with the NegaMax algorithm presented in Chap. 3. A comparison of the two algorithms is worthwhile. Perhaps it is useful for the learning effect if you create a copy of the NegaMax agent (context menu -> Duplicate...) and manually transform it into the rollout variant for practice. Very little needs to be adjusted for this. The difference between the two algorithms is that when simulating the opponent, the value is not recursively fetched for all possible opponent actions, but only one random action is selected at a time. The tree is thus only traversed on a single random path from the root directly to a leaf, i.e., without the jumps back after "working off" a node, which is characteristic for depth-first search.

Finally, the results of the numerous rollout simulations must be statistically evaluated. This is done by a higher-level function that starts the simulations in a counting loop and collects and evaluates the returns.

recursive "Monte Carlo Rollout" for TicTacToe

```
public double evaluateAction( int action, char player ){
  double sum = 0;
for (int i=1;i<=samplesNumber;i++){
    double v = rollout_evaluation(action,player);
    sum+= v;
```

```
        return sum/samplesNumber; }
}
public double rollout_evaluation( int action, char player ){
  state[action]=player; double reward = reward(state, player); if
(reward!=0) {
        state[action]='-';
        return reward; }
  // opponent player = (player=='o') ? 'x':'o'; ArrayList <Integer> A =
coursesOfAction(state);
if (A.size()==0){
        state[action]='-'
        return 0; // board full } double value = -rollout_evaluation(
                A.get(random.nextInt(A.size()), player );
state[action]='-'; // undo simulated action
return value;
}
```

It is remarkable how quickly a relatively high evaluation quality can be achieved with this relatively simple and random approach. In the following, we would like to find out if and when the MC rollout tactic generates an optimal TicTacToe policy. To do this, we gradually increase the number of rollouts and observe success against an optimal "NegaMax" policy. Here, we first let the rollout algorithm start in the "offensive" position as a beginning X-player.

In the class "TicTacToe_Environment" of the scenario "Chapter 4 Decision-Making and Learning\TicTacToeWith_Q_MCRollout_MCTS," a routine is prepared with which you can perform comparison tests. You can find the corresponding method in the code of the class under the declaration.

```
public   void   agentCompare(int   games,int   parameterMin,   int
parameterMax,
            int parameterStep)
```

To start the function in Greenfoot, you have to select it in the "context menu of the world." To do this, you need to click in the area between the playing field and the edge of the window (cf. ● Fig. 4.19).

With this function you can let the set agents play against each other and evaluate the results. You can set how many games the agents play against each other per iteration step. You can iterate over a certain parameter, in our case the number of rollouts. To do this, you first define the start and end value of the parameter and set the step size of the parameter change.

For games=1000, parameterMin=1, parameterMax=101, and parameterStep=1, you should get approximately a picture like in ● Fig. 4.20 if you let the "Monte Carlo rollout" play offensively, i.e., with X and "NegaMax" defensively, i.e., with O.

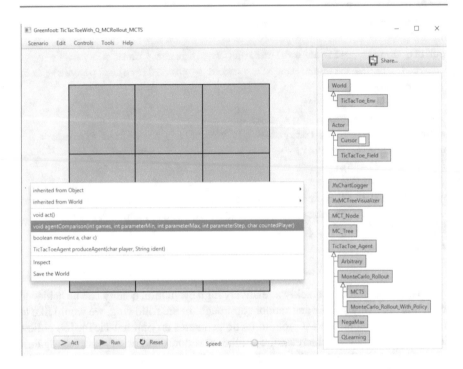

Fig. 4.19 In the "context menu of the world" you can start the routine "agentComparison"

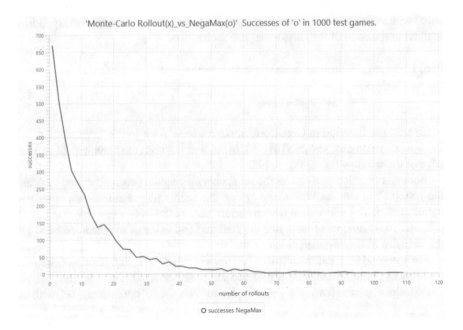

Fig. 4.20 Won games of a Monte Carlo rollout control against an optimal tactic (NegaMax algorithm) in the defensive starting position

```
public class TicTacToe_Env extends World{
    public final static String x_player = "Monte Carlo Rollout";
public final static String o_player = "NegaMax";
...
```

Make sure that you define in the function "agentCompare" from which player you want to display the won games in the chart. If the optimal policy "NegaMax" is to play a role, it makes sense to display the winnings of this player, since the opponent has no chance of winning (e.g., "o"). Values lower than 100% are caused by the fact that the opponent of the optimal policy sometimes reaches a "draw."

You can see that already with about 100 rollouts the optimal control does not win more than 99% of the games. NegaMax plays here from the defensive position, i.e., the optimal tactic is played by the player who bets second. With TicTacToe it is so that one always reaches a draw with optimal play.

A different picture emerges when NegaMax plays in the offensive position. Feel free to try it out by adjusting the algorithms used by the agents accordingly. It should then look something like the picture in ● Fig. 4.21 will result.

If the rollout algorithm is in the defensive position, then it does not reach parity even with 1000 rollouts; if the offensive player plays with optimal tactics, then the rollout algorithm practically never loses much less around 40% of the games. What is the reason for this?

If you take a closer look at the plays that the rollout tactic loses, they are exactly the ones that are commonly referred to as "quandaries" or "traps." It's worth noting that we've created a mechanism that filters "interesting" plays for us, so to speak.

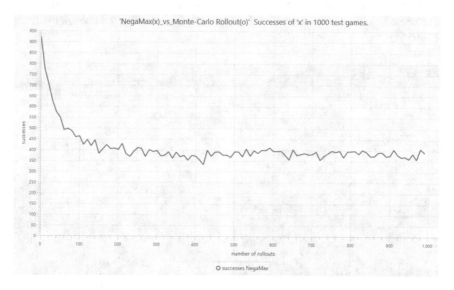

Fig. 4.21 In the defensive position, a purely random rollout at TicTacToe does not become equal to the NegaMax algorithm

In TicTacToe, a somewhat greater "intelligence" is apparently required of the defensive player than of the offensively playing agent in order not to fall into such a "trap." However, there seems to be a kind of "intelligence barrier" in the purely random rollout, because even by greater computational effort, the algorithm does not manage to master this challenge. What is the reason for this?

For this we can take a closer look at how the rollout algorithm reacts to such a "trap." To do this, we open the code of the class TicTacToe_Environment in the scenario "TicTacToe with MonteCarlo" and replace the specification "NegaMax" with "Human" for the x_player. In addition, we prepare a trap as the starting position.

```
public final static String x_player = "Human";
public final static String o_player = "Monte Carlo Rollout";
protected char[] matrix ={    '-','-','-',
                  '-','o','-',
                  'x','-','-'};
```

After starting with "Reset" and "Run" (if necessary, the TicTacToe_environment must be created beforehand by selecting the constructor), we place our cross in the upper right corner. To ward off the impending "pinch," the algorithm should actually set in one of the middle edge fields, but instead the algorithm sets in a corner in the vast majority of cases, allowing us to complete the trap. "NegaMax" recognizes the problem and evaluates correctly; ● Fig. 4.22 shows the different state estimates.

The rollout tactic exhibits a very "naive" gameplay. It is true that we could increase the success rate by integrating prior knowledge or tricks, e.g., by recognizing "traps" or evaluating drawn games against the optimal policy as a "success." The naive behavior of the algorithm becomes explicable when we consider that the

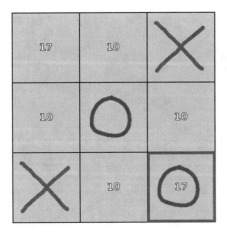 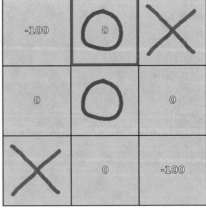

Fig. 4.22 The rollout policy on the left cannot detect the impending trap even after 100,000 evaluations, in contrast to the "NegaMax" algorithm (right)

move in ◉ Fig. 4.22 executed against a randomly playing opponent is often reward-ing. The reality, however, is different.

4.3.3 Artificial Curiosity

In more complex environments, we are dependent on optimizing exploratory behav-ior and primarily seeking "exciting" states during exploration, i.e., states that we do not yet know and in which a great deal of new knowledge can be acquired. There are several "intuitively" comprehensible approaches for the optimization of exploration.

So far, the behavior of our agent has been determined only by its policy and the specific reward function to which this policy responds and which we maximize in the learning process. Are there goals that—in view of complex state spaces with partly unpredictable subsequent states and only sparsely available rewards—are worth pursuing?

Energy and time consumption have existential importance in "real-world" con-texts. Therefore, we should consider not only seeking "exciting" states but also avoiding unnecessary repetitive actions and states. What conclusions do we draw from this?

The obvious thing to do is to check whether we already know a current observa-tion. In doing so, we want our agent to be "excited" when it makes a new observation and "bored" when it already knows an observation. Therefore, we now provide a kind of "curiosity module" that generates an "intrinsic reward signal" and adds it to the "extrinsic reward" from the environment. The policy is now trained to optimize the sum of the extrinsic and intrinsic reward signals. The setup was inspired by the work (Pathak et al. 2017) inspired, although we will only be concerned with basic principles here. We will also still provide for a "boredom" signal to ensure a "decay" of the "curiosity rewards" and the use of actions not yet chosen. The "boredom" acts in a similar way as the transition effort set in the RL_GridEnv_FV class (Fig. 4.23).

In this way, frequently visited states are increasingly evaluated negatively, and new "territories" at the edge of the known range, where many discoveries are made, are evaluated positively and appear correspondingly attractive.

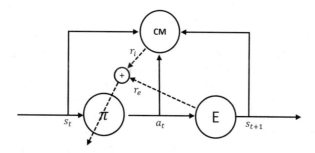

Fig. 4.23 The policy is trained to optimize the sum of the extrinsic reward and the boredom/curiosity-based intrinsic reward

For basal experiments, the relatively straightforward Q algorithm is often used. At this point, however, it does not work as well because it does not respond well to dynamic changes in the neighborhood of states. With Q, an action score only changes when the corresponding maximum in the subsequent state changes, which is why we will switch to the on-policy SARSA approach for the following experiment.

For the experiment, the model used in connection with the Dyna-Q was $model(s_t, a_t)$ was further developed so that it can also store several subsequent states. The intrinsic reward is now generated in a very simple way during the update: If the observed subsequent state was unknown, the update function returns the "curiosity" reward, and if the observation was already stored, the "boredom" reward is returned. In our case, the "curiosity" reward is in the order of magnitude of the possible positive rewards, while in contrast the "boredom" reward was chosen to be much smaller. This has the effect that he can also let go of discovered treasures in order to devote himself to curious exploration.

```
public final static double REWARD_INTRINSIC_CURIOSITY = 1;
public final static double REWARD_INTRINSIC_BOREDOM = -0.02;
```

You can find the implementations in the scenario "HamsterWithPolicyGradient_curiosity." For this, the classes implemented for ▶ section "Feature vectors and partially observable environments" feature vectors and POMDP) were supplemented accordingly. The map "mapFeatureFlat1" used is largely similar to the map "mapFlat," but the "rooms" now have walls with different textures to give the agent additional orientation options and occasions for "curious behavior." The learning curve of the SARSA agent with the "curiosity module" shows a higher volatility in comparison; however, it is clearly visible how now even more distant rewards in the environment are safely discovered. The curves shown in Fig. 4.24 are the learning progress (reward collected per episode) of twice the same algorithm but once with "curiosity module" and once without.

The implementation still leaves a lot of room for further improvements and investigations, e.g., the better consideration of the stochastic properties of the POMDP environment or by an augmentation of observations.

It is also exciting to look at the systematic exploration of the SARSA algorithm when the exploration parameter ε is set to 0. In the simple case, without a curiosity module, it becomes evident how the transition cost causes actions that have been performed to be negatively marked and the agent then chooses the actions that have not yet been performed in each case. Because of the agent often returns to previous states, either by own movements or the resetting to the start state, this is a bit reminiscent of the breadth-first search approach. Positive rewards now prompt the agent to purposefully go into depth as well. Curiosity rewards "encourage" the agent to abandon explored rewards and also to venture to the edges of the known area. This happens mainly when the valuations complemented by intrinsic rewards, exceed the available valuations generated by extrinsic rewards. One approach that systematically addresses this balance of "breadth" and "depth," as well as "exploration" and "exploitation," is Monte Carlo tree search, which is introduced in ▶ Sect. 4.3.4.

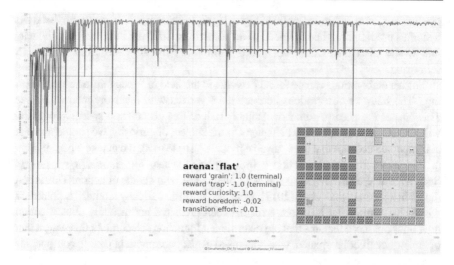

Fig. 4.24 Comparison of the learning curves of SARSA without (red) and with curiosity module (green) in a POMDP environment (feature vectors). Both agents otherwise have exactly the same learning parameters

Further Considerations

The topic of "artificial curiosity" has many exciting cognitive science references to questions of self-supervised generation of relevant features and categories. The notion of predictive behavior also comes from the field of cybernetics and systems theory (cf. Chap. 6). This is related to the model of "predictive coding" from neuroscience, where in principle a system component is assumed to act in such a way as to minimize a prediction error. Here, a model of the surrounding system is created that can estimate what consequence an action will have, i.e., what state (or reward) will be a will be achieved $model : (s_t, a_t) \rightarrow \hat{s}_{t+1}$. This type of model is also called a "forward model." This, very similar to the principle presented above, gives us a value that we can identify with a degree of "surprise." If we identify "surprise" with "curiosity," then our "curiosity signal" would consist of the distance $\Delta(s_{t+1}, \hat{s}_{t+1})$ between estimated and actually observed subsequent state.

However, it has also been shown that this criterion introduces new problems in complex and dynamic environments: Any kind of novelty becomes attractive, regardless of whether it is in our interest. In this context, the "noisy TV problem" is often cited, where an agent freezes in front of the white noise of a TV with no reception because every state on the screen is completely unpredictable and would thus be extraordinarily exciting for him. Not every kind of novelty also triggers "curiosity" in us. Pure unpredictability is not particularly exciting in the long run.

Let us take a closer look at the trade-off between the need for security and curiosity. If we want to gain security with regard to our behavior, we strive not only for observations to meet our expectations, but also, in principle, to predict our own behavior. This leads to our behavior quickly becoming stuck in early well-trodden paths and, moreover, becoming increasingly deterministic and predictable. We would now like to consider as "boring" all actions and states in which we can predict our own behavior with a high degree of certainty. If our own behavior becomes predictable, then the need to do something different increases. By the way, this would also mean that we would soon find "freezing" in front of the TV set boring.

Further elaborations can be found in works in the field of "deep reinforcement learning." The states or observations here are large vectors with relatively many elements. This creates the need to somehow suitably reduce this observation space, a topic that will also concern us in the next chapter. On the subject of curiosity, we should mention at this point a concept that uses "inverse models." These models can take a pair of states (s_t, s_{t+1}), the corresponding action a_t from a pair of states, $inv\_model\left(s_t, s_{t+1}\right) \rightarrow \widehat{a_t}$. Such a module allows us, so to speak, to predict our own choice of action. This is also the main topic in Pathak et al. (2017). Here, an "intrinsic curiosity module" is presented, which, by minimizing thRussee error between predicted and actually chosen action, now finds out those features that are determinant for action selection. In this way, a kind of "relevance filter" is created, which learns to neglect unimportant pattern components.

Reliability of Statistical Estimates

We can assume that the exploration of states that we have already visited frequently is less worthwhile than the exploration of states that are still unknown. Imagine two of the well-known "one-armed bandits." Both gambling machines have a certain probability of winning, but the probability is unknown to us. If we pull the lever once or twice, we cannot yet say much about the probability of winning on the respective machine. We can be "lucky" and win on the—in the long run—unfavorable slot machine, or we can initially lose on the machine with a higher probability of winning. In these cases, we would get a "wrong impression" of the machines. However, our stochastic estimate gets better the more often we pull the levers. Although there are common stochastic methods to estimate the confidence of samples, to treat the topic exhaustively in our context is not trivial at all, since we are also dealing with the consideration of structural properties of a largely unknown environmental system.

It is often obvious to first keep statistics about the states visited in the past and the actions selected in them. For this purpose we create, e.g., a table $N(s,a)$ in which it is counted how often state-action pairs have already been visited. Now we can, e.g., calculate the learning rate η,e.g., with $\eta = 1/N(s,a)$ or the exploration rate can be made dependent on these visit statistics.

We can also develop a policy in which we tend to avoid states that have been visited very often, not only greedily deciding according to the largest observed $\hat{Q}_t\left(s,a\right)$-value but additionally adding a kind of intrinsic reward $U_t(s,a)$, which is inversely proportional to how many times the action a in s has already been performed. The agent would thus decide according to the maximization of $\hat{Q}_t\left(s,a\right) + U_t\left(s,a\right)$ decide, too. The effect of this is that the behavior is additionally driven by improving the reliability of the profit estimate. The so-called "upper confidence bound" has proven to be favorable for this. This principle is used in (Monte Carlo tree search). A more detailed consideration of this method can be found in the next section.

4.3.4   Monte Carlo Tree Search (MCTS)

If we want to simulate and weigh our decisions in a more elaborate way, then we have to deal with a rapidly growing complexity and the associated computational effort. Another very interesting algorithm that builds on the "rollout strategy" is the Monte Carlo tree search. MCTS has also played a central role in the spectacular successes of computer Go.

As mentioned earlier, if we produce all possible future world states, we quickly get a huge explosion in the size of the state space. We generally do not have the time or computational power to search the tree of possible subsequent states sufficiently well for an optimal solution. This is a problem we encounter even in board game scenarios, even though we are dealing with comparatively very simple, discrete, and controllable environments.

If the calculation is terminated prematurely after a certain period of time, two central problems arise: Firstly, the area searched up to that point may not yet have been correctly evaluated, and secondly, there are still large areas in the state space that have not been searched at all.

MCTS addresses both of these challenges: Within the known space, it makes sense to primarily follow the most successful paths so far, although it is also necessary to note that the agent tries new possibilities every now and then to perhaps discover better paths. It is true that these measures already allow us to reduce the state space to be checked enormously. However, at some point we reach a state where we leave the known space. This is where MCTS expands the "game tree" by adding another node to it. Outside of the entered and therefore not evaluated area, we can now actually only decide arbitrarily or heuristically in our model along possibly recognized "landmarks."

The idea with MCTS is to have a "rollout" at this point (cf. ▶ Sect. 4.3.2), without processing the simulated experiences in detail. This takes into account that the cost of storage space and computation time used is relatively low with respect to the benefit and quality of the information obtained. The ultimate success or failure of this rollout is then entered at the state that was added to the tree. The simulated episode, the "sample," is thus initially used only to estimate the value of the last "leaf."

Finally, the new "virtual experience" is retroactively processed within the evaluation tree by adjusting a "success statistic" for each node above it—the new information is "backpropagated" in the tree, so to speak.

MCTS is therefore divided into four phases ◉ (Fig. 4.25):

1. Selection
2. Expansion
3. Simulation
4. Backpropagation

These steps are repeated as long as there is still time available. It is a very interesting and extremely practical property of the algorithm that at any time we have a

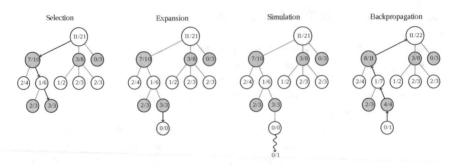

Fig. 4.25 Phases of a run in "Monte Carlo tree search" (Mciura, Dicksonlaw583 [CC BY-SA 4.0] from https://commons.wikimedia.org/)

more or less usable, preliminary result, the quality of which depends on the computing time used at the time. MCTS therefore belongs to the "anytime algorithms."

At the beginning of the evaluation, there is only one root node in the tree. It corresponds to the state in which the agent is currently located. With each run, the game
tree grows by another node. The four phases of a run in MCTS are described in
detail below.

Step 1— Selection

First, we must move within the game tree to the state we wish to evolve. In
MCTS, the game tree is not evolved uniformly by a "blind" systematic, as in a
breadth-first search, for example. In MCTS, a "success statistic" is stored in each
node of the game tree, indicating how many simulations have already been started
from that node and how many of them ended successfully. The tree grows asym-
metrically along successful paths, as we follow tactics within the tree that are—to
some degree—greedy. This tactic within the tree, which has to balance the conflict-
ing requirements of "exploration" and "exploitation," is called "TreePolicy."

Spontaneously, one might think that it makes sense to simply use the average
success of an action a by dividing the sum of the episodes that were successful w_a
by the number of simulations performed.

$$\overline{R}_a = \frac{w_a}{N_a}$$

However, this is only true to a very limited extent, as such a sample would only
represent a point estimate from a stochastic point of view. The larger the sample, the
more reliable such an estimate becomes. To illustrate, consider an example. Imagine
that in a situation s we are given two cubes a_1 and a_2; cube a_1 has two golden colored
sides, and cube a has $_2$only one. We gain G = 1 if one golden side is facing up. We
now need to find out empirically which cube is the better one. If we do only a small
number of trials, it can easily happen that we are unlucky and the "bad" die a wins
by $_2$chance more often than the "better" die a_1, and thus we finally choose the wrong
one. How reliable are these estimates?

Suppose we get a total of 15 wins with$_1$ a after $N_{a1} = 50$ tries, we get a total of 15 wins and with a $_2$after $N_{a2} = 10$ rolls we get 3 wins. This would mean:

$$\overline{R_1} = 0{,}3 \quad \text{and} \quad \overline{R_2} = 0{,}3$$

Intuitively, it is clear that the estimate $\overline{R_1}$ is much safer than the estimate $\overline{R_2}$. To answer this question quantitatively, the so-called confidence intervals are used.

For the printout:

$$c_{N,N_a} = \sqrt{\frac{2 \ln N}{N_a}}$$

which is based only on the number N_a of trials performed with a and the total number N of all trials; it is shown that the probabilities:

$$P\left(\overline{R_1} \geq \mu_a + c_{N,N_a}\right) \quad \text{and} \quad P\left(\overline{R_1} \leq \mu_a - c_{N,N_a}\right) \quad \text{are smaller than} \quad \frac{1}{N^4} \quad \text{("Hoeffding}$$

inequality").

It can be seen from the inequality that as the number of trials N increases, the uncertainty decreases very rapidly. Here μ_a represents the theoretical, "real" expected value that would be obtained after an infinite number of trials. For the cubes a_1 and a_2, for example, one third and one sixth).

The UCB1 formula (for "upper confidence bound") uses this upper interval bound for evaluation. The transfer of the UCB1 formula to a "Monte Carlo tree" was first presented by Kocsis and Szepesvári at the ECML 2006 in Berlin (Kocsis and Szepesvári 2006):

$$UCB1\left(\overline{R_a}, N_a, N\right) = \overline{R_a} + c_{N_a,N}$$

Kocsis and Szepesvári recommended to select the node a within the tree for which the value of:

$$\frac{w_a}{N_a} + c\sqrt{\frac{\ln N}{N_a}}$$

is maximum. Thereby w_a the number of successfully performed simulations and N_a is the number of all trials of a performed. N represents the total number of trials of the parent node and c is an "exploration parameter" which theoretically has the value $\sqrt{2}$ but in practice is often determined empirically.

In the example, we obtained for the action option cube $_1$ am with UCB1a value of $\approx 0{,} 7$ and for the action option cube aa_2 value of$\approx 1{,} 2$. UCB1 would thus tend to select the more uncertain node with the smaller sample, given the same average score, improving its estimation. One can therefore interpret this, with some imagination, as a variant of artificial curiosity, in which the possible reduction of uncertainty is incorporated into the action evaluation. This exploratory component becomes smaller the more often the action has been performed. Not only is each action possibility evaluated individually, but in each case the context, i.e., the total number of samples performed, is also taken into account. This can also lead to nodes that previously had a success rate of 0 being selected from time to time if the number of its visits is comparatively small and the corresponding relative evaluation uncertainty is large.

Step 2—Expansion

If a node is reached that does not mark a terminal state and has not yet been fully expanded, one of the unknown actions is selected, and a corresponding node is appended to the game tree.

Step 3—Simulation

In this step, the "rollout" takes place, which in the original variant runs randomly and does not save the individual simulated experiences, whereby the only aim is to determine whether the rollout simulation lands a "hit" or is unsuccessful. However, there are also some extensions to this.

In so-called "real-time MCTS," the rollout is truncated after a certain depth or by an evaluation function. This can be useful in scenarios where no unique terminal states are available that are marked with "won" or "lost."

Furthermore, the simulations do not necessarily have to be purely random. In such approaches, the simulation actions are selected by a "rollout policy." Generalizing, we can then say that in MCTS we have two policies, a "tree policy" that is valid inside the developed tree and a "rollout policy"—which is often a "random policy"—that is applied outside the tree for controlling the rollout simulations.

Step 4—Baking Propagation

With the result of the rollout, the statistics must now be updated in all nodes on the path from the evaluated node to the root node.

A Java Implementation of the MCTS
For the implementation of MCTS in Java, we develop the rollout example in our Greenfoot scenario from ▶ Sect. 4.3.2 to make further use of the rollout features. However, since we no longer evaluate each action option by itself in MCTS, i.e., independently of the others, we need to overload the policy of the rollout in the MCTS agent. It is not enough to just implement a new evaluation function.

MCTS Policy

```
@Override
public int policy(char[] state){
  char[] backup = getState(); char opponent = (ownSign=='o') ?
'x':'o';
  if (TicTacToe_Agent.countActionOptions(state)==0) return -1;
  MC_Tree mct = new MC_Tree(state);
```

```
 MCT_Node root = mct.getRoot();
 root.setState(state);
 root.setPlayer(opponent);
 long endTime = System.currentTimeMillis()+timelimit; int c=0;
    while      ((System.currentTimeMillis()    <     endTime)     &&
(c<getMaxRollouts())){
    MCT_Node selectedNode = selection(root);
    if (TicTacToe_Environment.checkMatrixWon
(selectedNode.getState())=='-
'){
       expand(selectedNode);
    }
    MCT_Node nodeToBeEvaluated = selectedNode;
    if (selectedNode.getNumberOfChildren() > 0)
       nodeToBeEvaluated = selectedNode.selectChildRandomly();
    }
setState(nodeToBeEvaluated.getState());
          double   v   =rollout_evaluation(nodeToBeEvaluated.
getActionFromFather(),
          nodeToBeEvaluated.getPlayer());
    backpropagation(nodeToBeEvaluated, v);
    c++; }
 MCT_Node bestNode = root.childWithMaxScore();
setState(backup); return
bestNode.getActionFromFather();
 }
```

The four-phase sequence of selection, expansion, simulation, and backpropagation can be clearly seen in the implementation of the MCTS policy. A Java implementation of the functions that handle these phases can be found in the box below. For the rollout, we use the function inherited from the "MonteCarlo_Rollout" superclass.

MCTS Phases

```
private MCT_Node selection(MCT_Node root){
 MCT_Node node = root; while (node.getNumberOfChildren() != 0) {
    node = node.childWithMaxUCT();
}
 return node;
}
private void expand(MCT_Node node){ char[] fatherState = node.
getState();
char childNodePlayer = (node.getPlayer()=='o') ? 'x':'o'; List
<Integer>
possibleActions = coursesOfAction(fatherState); for (int a :
possibleActions){
```

```
      char[] childState = fatherState.clone();
      childState[a]=childNodePlayer; // perform action
      MCT_Node newNode = new MCT_Node(childState);
      newNode.setPlayer(childNodePlayer);
      newNode.setFather(node);
      node.addChild(newNode,a);
   }
}
private void backpropagation(MCT_Node node, double score{
  MCT_Node temp = node; score = Math.round(score); char winner
= '-'; if
(score>0) winner=node.getPlayer(); if (score<0) winner=node.
getOpponent();
if (score==0) winner=ownSign; while (temp != null) {
      temp.incVisits();
      if (temp.getPlayer() == winner) {
      temp.addScore(1);
      }
      temp = temp.getFather(); }
}
```

In our TicTacToe environment with an optimally playing opponent against whom we can never win, we evaluate a draw as a success for our agent. When we test run in the defensive position, we see that the MCTS agent loses virtually no games against the optimal policy from about 1700 samples ◉ (Fig. 4.26).

What is the superiority over the simple rollout algorithms? It is important that the tree policy within the developed game tree also simulates the behavior of the opponent, since we save successes of the opponent. Moreover, we continue to follow successful paths more intensively and deviate from a successful path only within the "confidence bounds."

This has parallels to the behavior of the Monte Carlo hamster in ▶ section "Subsequent evaluation of episodes ("Monte Carlo" method)" whose behavior is also primarily shaped by successful episodes. The procedure also allows us to master games with larger and "low-success" state spaces.

Some Remarks About Alpha Go Zero
Playing Go on a human level was long considered a challenge that is actually unsolvable by means of computer science because of the astronomical number of move variations, also in comparison with chess. This assumption was impressively disproved by Google DeepMind in March 2016 in the victory of the program AlphaGo against the Go world champion Lee Sedol.

The AlphaGo algorithm falls into the area of so-called "deep learning," which uses artificial "deep neural networks" to enable machine learning even in complex state or action spaces. In the context of this book, we can unfortunately only deal

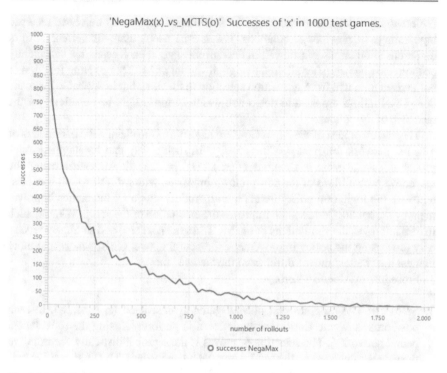

Fig. 4.26 TicTacToe games won by an optimal policy (out of 1000) against an MCTS policy.

with this topic in passing. However, we will briefly discuss neural networks and their applicability in the context of reinforcement learning in the next section.

Building on the experience with AlphaGo, a DeepMind team developed a "deep reinforcement learning" software called AlphaGo Zero (Silver and Huang 2016). Unlike AlphaGo, this program does not use any human-knowledge about the game. It learned exclusively from self-play reinforcement learning, where the input provided only descriptions of the placements of pieces on the Go board without any preprocessing. Another difference, in addition to not using human prior knowledge or human-designed features, is that AlphaGo Zero used only a single one of these deep neural networks (a "deep convolutional neural network") combined with a simplified version of Monte Carlo tree search (MCTS).

In AlphaGo Zero, a form of policy search was implemented in which policy-based state evaluation was interleaved with policy-improvement. A structural difference between the algorithm of AlphaGo Zero and AlphaGo is that AlphaGo Zero used a Monte Carlo tree search to select moves during self-play reinforcement learning, while AlphaGo applied Monte Carlo search only for a subsequent improvement of live play but not during learning.

The MCTS generates a policy improvement (cf. ▶ Sect. 3.2.1) on the basis of the neural network's evaluations f_θ, by discovering somewhat improved moves through its exploratory rollouts, which are at the same time guided by success. Since MCTS preferentially selects and extends initial sections of trajectories that have received high ratings in previous simulations, it can thus accumulate "experience" from numerous simulations, starting from the current state in each case, in a sufficiently economical manner. After all, in the original variant of MCTS, no rating function or policy is maintained from one action selection to the next, but in AlphaGo Zero, the deep CNN ratings are used to determine which actions might be beneficial for the next step of the rollout.

The inventive idea of the AlphaGo-Zero algorithm is to change the parameters of a neural network (cf. next ▶ Chap. 5) $f_\theta(s)$ in such a way that its supplied move probabilities and state evaluations $(P(s, .), V(s)) := f_\theta(s)$ fit somewhat better the respective probabilities of the new policy π improved with MCTS, which is consistent with the logic of a policy iteration with "model support": the tactics improvement by the Monte Carlo search improves the estimation of the action values, which in turn improves the simulations of the MCTS in the next step and so on. By repeatedly processing the policy improvement, the neural network learns more and more in what the "better" moves differ from the "worse" moves.

Procedure at AlphaGo Zero:

1. Self-Play: The program first plays a game against itself $(s_1, s_2, ..., s_T)$. In each position s_t a Monte Carlo tree exploration is performed using the most recent neural network f_θ. The selection of moves is done probabilistically according to the values calculated by the MCTS in π_t calculated values. The final position s_T is evaluated according to the usual rules of the game.

2. Network Training: The neural network takes the non-preprocessed board position s_t as input, passes it through many convolutional layers with the parameters θ and outputs both a vector p_t which represents a probability distribution over the possible moves and a single score v_t which represents the probability that the current player with the position s_t wins. The update of the neural network parameters θ now aims to improve the similarity of the vector p_t with the improved search probabilities π_t from the MCTS exploration, as well as to minimize the error between the predicted winner v_t and the actual game winner. The new parameters θ will be used in the next iteration of Self-Play (Step 1).

Alpha Go Zero thus combines retrospective processing of experience ("neural network"-based policy iteration) with predictive simulations (Monte Carlo tree search). This is certainly the reason for the spectacular success and groundbreaking properties of the algorithm.

4.3.5 Remarks on the Concept of Intelligence

"Model-based methods" today tend to be attributed to an area of AI that is not counted as "machine learning" in the strict sense, since "model-free methods" have emerged, roughly speaking, in contrast to and in distinction from the "good old-fashioned AI approach." GOFAI was about generating (symbolic) representations of the world and searching and planning within that model. GOFAI had largely supplanted 'model-free methods' by the time of the crisis and renaissance in the 1980s (cf. Chap. 6). There had been clear barriers to the GOFAI approach.

The "intelligence" of an agent system with specific sensorimotor preconditions is shown by how diverse and extensive the conditions are under which a system is allowed to act in a meaningful and goal-directed manner. Thus, intelligence is not first about managing a model's approximation to a given "truth" but about producing successful behavior. Internal models, therefore, need not reflect a "pure truth" but must first and foremost represent useful images that allow us to plan meaningfully and test "hypotheses." In constructivist terms, there is no such thing as "partially observable reality"; there are only useful and less useful models. The crucial question is: Is the model able to predict well what is currently happening, or what would happen if?

This view touches on the concept of "truth." Does "objective truth" hereby become meaningless? It is true that it does not become meaningless, but the concept is "relativized." "True," this is hereby no longer an attribute that attaches to any concepts or statements "objectively." Rather, it indicates that a statement is comprehensible to a receiving agent and therefore useful to it. "Truth" thus becomes an expedient term within a social or a "multi-agent" context, by which we tell the receiver that it can incorporate a piece of information into its own portfolio in a specific way that is likely to increase its ability to act meaningfully.

"Skinnerian creatures," to use the terms of D.C. Dennett (2017) are purely reactive creatures endowed with "competence without understanding," such as a moth that flies against a lamp to the point of exhaustion and is unable to see through its properties, or a spider that weaves highly competent elaborate fishing webs but probably does not even begin to understand what it is actually doing and is thus also forced to do so again and again in the same way, given the conditions that trigger the behavior. A quantitative adaptation of this "skill" can only fundamentally change this by accidental circumstances and only in a protracted "unreflective" process.

Planning action also holds the potential to take our agent's behavior to a higher level. The "virtual goal search" in the model allows to *explicitly* determine goals and thus also reasons for action, which is a necessary precondition for leaving the stage of "Skinnerian creatures." The answer to each "Why?" is a motivation provided by goal-directed modeling.

We have, when looking at the behavior of the Monte Carlo rollout agent, justifiably spoken of NegaMax (with its optimal policy) behaving "more intelligently" than the rollout algorithm. This again points us to the applied notion of "intelligence." Let us therefore venture a consideration of the applied intelligence measure: Given a given environmental system and equal worthwhile goals therein, as well as equal action possibilities in each case, the "more intelligent" or "more skilled" agent can successfully handle a larger set of given world states. For probabilistically decisive agents, we can say that the average cumulative rewards achieved for a larger set of states are larger for the "more intelligent" than for the "less intelligent" agent.

Shane Legg and Marcus Hutter 2007 formulated a definition for "intelligence":

Intelligence measures an agent's ability to achieve goals in a wide range of environments.

In their paper, the authors have condensed 70 different definitions of intelligence into this phrase. In their definition we find the two mentioned factors "ability to achieve goals" and "wide range of environments."

With this concept of "intelligence"—which is still very limited, e.g., there is no social, environment-creating aspect—intelligence would depend on two factors: Firstly, the number of situations in which an agent is able to act meaningfully and secondly, the average success that he achieves in these situations, considered over his period of action. The speed of adaptation to new environments also represents an important variable here, because the gain is very small at the beginning, compared to what is cashed in later. This becomes especially important if the agent's flexibility is to be great.

Compared to, e.g., human intelligence, a TicTacToe or even a chess agent is still very stupid, because the set of environmental states in which it can act successfully is limited to TicTacToe or chess playing fields.

4.4   Systematics of the Learning Methods

At this point we can try to gain a more general overview of the components of a learning system and their relationships and roughly assign the presented procedures and algorithms to the relationships between the components and place them in a larger context ◉ (Fig. 4.27).

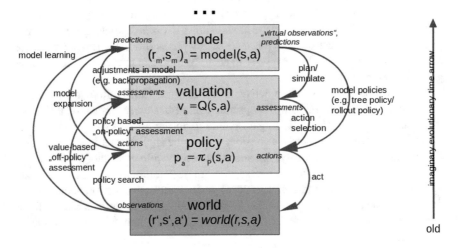

Fig. 4.27 Systematics of learning components and methods

The rough arrangement of the system components shows a certain plausibility with regard to an "imaginary evolutionary development." This "imaginary" time arrow serves to add an additional dimension to the representation, a kind of imaginary "earlier" and "later." It thus serves as a structural overview and does not correspond to a "really" expired evolution.

The "oldest" element here is certainly the policy, as the component that directly assigns actions to sensory states. After that comes the component that allows state evaluations and thus goal-oriented decisions, while a model enables the simulation of actions and can be used to "virtually" identify goals and optimize paths, which can also improve the state evaluations and the policy.

Certainly questions are raised that are not dealt with within the scope of this book, much is also still the subject of current research and may inspire the reader to do their own research.

The algorithms presented so far could technically only work through two massive simplifications. First, the environments they interacted with were limited to a more or less manageable number of states; second, all worlds were of a clearly defined type in terms of their dynamics and interaction possibilities (gridworlds and game boards).

If we want to significantly increase the states of the environmental system, we quickly reach limits with our explicit representations in the form of tables or trees. In the following chapter, we will address this problem.

Bibliography

Dennett DC (2017) From bacteria to Bach and back: the evolution of minds. Norton, New York

Kocsis L, Szepesvári C (2006) Bandit based Monte-Carlo planning. In: Fürnkranz J, Scheffer T, Spiliopoulou M (eds) Machine learning: ECML 2006. 17th European conference on machine learning, Berlin, Germany, September 18–22, 2006; proceedings. ECML; European conference on machine learning. Springer, Berlin (Lecture notes in computer science Lecture notes in artificial intelligence, 4212), pp 282–293

Godfrey-Smith P (2019) The octopus, the sea and the deep origins of consciousness, 1st edn. Matthes & Seitz Berlin, Berlin

Legg S, Hutter M. A collection of definitions of intelligence. In: IDSIA-07-07

Mnih, Volodymyr, Adrià Puigdomènech Badia, Mehdi Mirza, Alex Graves, Timothy P. Lillicrap, Tim Harley, David Silver, und Koray Kavukcuoglu. Asynchronous Methods for Deep Reinforcement Learning", 2016

Pathak D, Agrawal P, Efros AA, Darrell T (2017) Curiosity-driven exploration by self-supervised prediction. arXiv:1705.05363

Russell S, Norvig P (2010) Artificial intelligence. A modern approach, 3rd edn. Pearson Education, Inc., New Jersey

Silver D, Huang A (2016) Mastering the game of Go with deep neural networks and tree search. Nature. Available online at https://www.nature.com/articles/nature16961

Sutton RS, Barto A (2018) Reinforcement learning. An introduction, 2nd edn. The MIT Press (Adaptive computation and machine learning), Cambridge, MA/London

Williams RJ (1992) Simple statistical gradient-following algorithms for connectionist reinforcement learning. Machine Learning 8:229–256

Artificial Neural Networks as Estimators for State Values and the Action Selection

5

> *Just as feathers are irrelevant to flight, in time we may discover that neurons and synapses are irrelevant to intelligence.*
>
> Ethem Alpaydin (2019)

Abstract

Usually, the available resources are not sufficient to tabulate policy, valuation function, or model. Therefore, this chapter introduces parameterized estimators that allow us, for example, to estimate the valuation of states even if they have not been observed in exactly the same form before. In particular, the so-called artificial neural networks are discussed. We will also learn possibilities to use such estimators to create parameterized policies which, for a given state, can produce and improve a useful probability distribution over the available actions.

Supplementary Information The online version contains supplementary material available at [https://doi.org/10.1007/978-3-031-09030-1_5].

Since in very many cases it is not practical to store probabilities or scores for each state and action in a table, we need methods to calculate this using parameterized estimators.

These estimators have a large number of parameters w which are set to compute the desired values under the state (action) specification. Usually, however, the size of the parameter space is much smaller than the state space because the estimators can interpolate or extrapolate values. In certain designs, for example, the need to divide continuous state spaces into discrete intervals is also eliminated if we want to design adaptive controllers that are to operate in a continuous state space, for example, think of self-driving cars or autopilots. There is also another crucial advantage of parameterized estimators: they can compute meaningful values not only for the training set but also for states that have never been seen by the agent before. Whether this "generalization" works well, however, depends on whether the estimator has been chosen appropriately.

Supervised and unsupervised learning methods often fill the bulk of works on machine learning. A special attention is often given to function approximators based on neural network simulation. This variant of nonlinear function estimators has many interesting properties, also stimulates the imagination, and has a lot of impressive results to show. It is not at all self-evident that computers are capable of estimating a function like this: $cat : \mathbb{R}^{N \times M} \to [0; 1]$, which computes for an arbitrary matrix of brightness values whether it is a cat image.

Although supervised learners are not the main subject of this book, a small excursion into the subgroup of artificial neural networks should not be left out at this point. In the following, we will once again look at how they work and their most important properties and perform the famous "Hello world!" of (supervised) machine learning, the recognition of handwritten digits, and then attempt the regression of a financial market curve and extrapolate some values.

5.1 Artificial Neural Networks

The performance of biological cognitive apparatuses, especially the human brain, sometimes clearly exceeds the possibilities of the best engineering solutions. In the many historical attempts to generate "intelligence" artificially, it generally became very clear how unrealistic the ideas were. It turned out that not only the biological body of living beings but also the nervous system, including the brain, represents an extremely complex hardware implementation that has evolved over the course of tribal and individual history. It has also been recognized that cognition is less about logical thinking and explicit, symbolic knowledge, and more about (social) action and the independent processing of appropriate interactions. Our body with its highly developed nervous system "merely" enables us to behave successfully within our physical, biological, ecological, or social contexts, in our "milieus."

Logic and mathematics, on the other hand, are techniques that are not directly concerned with successful action but are auxiliary sciences that, with their rigorous models, help to accomplish a wide variety of practical, technical, or scientific tasks. They allow various questions to be answered computationally. The computer emerged from the effort to automate computation, also in view of military or economic challenges. Obviously, mechanized computation with computers is not related to cognitive processes in nature. A characteristic feature of evolutionary processes, however, is the leapfrog development through misappropriation of gradually evolved organs and instruments. Due to the proven ability of computers to reproduce all computable processes, it is theoretically possible to model more or less isomorphic replicas of all possible systems, given the appropriate resources, including weather models, nuclear reactions, or neural networks, including their unpredictable, "chaotic" evolutionary properties. It is certainly one of the most exciting discoveries of the twentieth century that in general deterministic, exactly determined, "calculating" systems ("Turing machines") can also have unpredictable properties (Turing 1937).

In view of the enormous amount of "not-knowing," it was obvious to first draw inspiration from the structure of biological models when constructing "intelligent" systems. This is similar to the approach taken in aviation, where birds were first started to be recreated with wood and canvas before the laws of aerodynamics could be discovered. Therefore, modern computer history has been accompanied by research on artificial neural networks from the very beginning. For example, McCulloch and Pitts described the first mathematical model for an artificial neuron as early as 1943. However, the biological hardware that generates the various cognitive outputs differs significantly from that of a common serial computer with von Neumann architecture.

While reinforcement learning produces some spectacular results, we do not actually yet possess a true "computational theory of intelligence," that is, a general theory of how information (stimuli) from arbitrary environmental systems is processed in such a way that we can act increasingly competently within them. If we possessed this, then perhaps solutions could be constructed, similar to aircraft, that are optimally tailored to the capabilities of the available silicon and metal-based hardware. In aeronautics, for example, it has been recognized that it is possible to dispense completely with the replication of springs (Alpaydin 2019). However, it is already the case today that artificial neurons actually have very little in common with their natural models.

The structure of the human brain is extraordinarily complex. A wide variety of systems and complexes interact at different levels of organization. For the most part, processing occurs in parallel. The most conspicuous elements of the brain are the neurons. Their number is estimated at about 10^{11}. These neurons operate in the millisecond range, which is actually relatively slow. What makes for the enormous "computing power" is the internal connectivity: a neuron in the central nervous system is connected via so-called synapses to an average of about 10^4 other neurons, all of which operate in parallel. The parallel architecture allows the brain to perform about 10^{13} to 10^{16} analog computational operations per second with a maximum chemical power consumption of only about 100 watts. The power consumption may be relatively high for a biological organ, but from a technical point of view, it is sensationally low: analog 10^{16} computing operations per second—this corresponds approximately to the computing power of IBM's Summit supercomputer (OLCF-4), which achieves up to $2-10^{17}$ floating point operations per second but because of its power consumption of about 13 megawatts almost needs its own power plant and weighs quite a few tons.

The basic structure of a neuron ◉ (Fig. 5.1) is described many times in the popular literature: a typical mammalian neuron is composed of dendrites, the cell body, and an axon. This cell process can be very long and thus enable excitation conduction over long distances. For this purpose, an electrical signal runs through the axon, which is generated by the targeted passage of certain ions through the cell membrane. The end of the axon is connected to other nerve cells or recipient cells, e.g.,

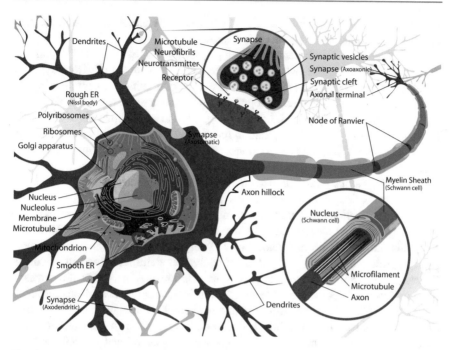

Fig. 5.1 Structure of a vertebrate nerve cell. (Author: "LadyofHats," published in the public domain on Wikipedia)

a neuromuscular end plate, via synapses, at which the signal is usually transmitted chemically.

Functionally, neurons can be divided into motor neurons, sensory neurons, and interneurons. Sensory neurons—also referred to elsewhere as afferent neurons or nerves—transmit information from the receptors of the internal organs or sensory organs to the brain, spinal cord, or nerve centers of the digestive tract. Motor neurons[1] transmit impulses from the brain and spinal cord to muscles or glands, where they trigger the release of hormones or cause muscle cells to contract, for example. The interneurons make up the largest set of neurons in the nervous system and are not specifically sensory or motor. They have a mediating function. They process information in local circuits or convey signals over long distances between different parts of the body. A distinction is made between local and intersegmental interneurons. Studies in the 1990s showed that both the diversity of neurons and the complexity of their modes of action in the cerebral cortex had been greatly underestimated: "Thus, based on anatomical and immunocytic criteria, the number of subtypes of cortical neurons would probably be between 50 and 500." (Churchland and Sejnowski 1997, S.53).

In terms of their action, neurons can be divided into two classes: excitatory and inhibitory neurons. An excitatory signal increases the probability that postsynaptic

---

[1] Also known as "efferent neurons" or "motor neurons."

cells will fire, while an inhibitory signal decreases this probability. Some neurons also have a modulatory effect on other neurons.

The scenario of reinforcement learning with its adaptive artificial agents has many parallels to the situation of natural, biological "agents" in their ecosystems. What tasks are assigned to artificial neural networks in reinforcement learning? In reinforcement learning, pattern recognition is to be seen primarily under the aspect of appropriate action selection. On the one hand, the approximation of the action function (policy) $\pi_P(s, a)$, the evaluation function $V(s)$ or $Q(s, a)$, or an environmental model $model(s, a)$ should be mentioned. Models provide which subsequent states and, if applicable, which rewards can be achieved if the corresponding actions were selected. In robotics there are also so called "inverse models." $modell^{-1}(s, s') = a$ which return which action is necessary to reach a desired state.

All these functions give their results as a function of the observed state of the world s which is an element of the space of all possible states S states. However, since this can quickly become unmanageably large, as we have seen, the estimators must in principle perform recognition of equivalent states, independent of "insignificant" changes in perception, such as shift, rotation, and insignificant changes in color or volume. This is where the ability of artificial neural networks to automatically extract helpful features from training data comes into play. It should be critically noted, however, that although this sounds very good at first, because it is exactly what we want, this "ability" is to be viewed quite critically in Ribeiro et al. (2016). It is shown, among other things, that these "characteristics" are not always what we would use with common sense. In an example described there, where dogs were to be distinguished from wolves, wolves were classified on the basis of the snowy landscape surrounding them, because the training pictures also showed wolves in a wintry environment, but dogs never in the snow. With such a network, dogs in the snow would automatically become wolves.

5.1.1   Pattern Recognition with the Perceptron

Compared to the natural ones, the artificial neurons represent a massive simplification ◉ (Fig. 5.2). The simplest and most basic model is Rosenblatt's perceptron. If x is the vector containing the input activations and w_j a vector corresponding to the stored connection weights to the output unit y_j, then the incoming activation can be conceived as the inner product of x and w_j. The incoming signals are finally transferred via a transfer function to the "axon", i.e., to the weighted "stimulation" of the connected cells.

In the perceptron, sensory neurons first propagate activation directly to output neurons through their respective weighted connections. In the simplest case, the input to the output neuron is a weighted summation of the inputs from the sensory units (sigma neuron).

The perceptron can calculate a kind of matching measure between two patterns. In the case of a linear transfer function, we obtained the "cosine measure." This is a "similarity measure" that uses the angle between two vectors for comparison and

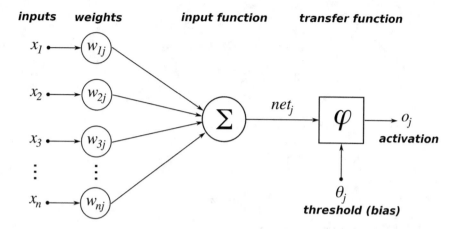

Fig. 5.2 Artificial "neuron". (Christoph Burgmer, license: GFDL (CC BY-SA 3.0). Source: Wikipedia)

thus can abstract from the magnitude of the vectors. If two vectors are congruent, the angle is 0 degrees, and thus the inner product is $x \circ w$, and thus the activation of the corresponding output unit is maximal. If one were to use the Euclidean distance according to a "Pythagorean theorem" generalized to higher dimensions, then, for example, there would be a corresponding distance between a lighter and a darker image on which the same subject is, and thus "dissimilarity" would be postulated. In cosine measure, on the other hand, vectors that have only been scaled, e.g., vectors (1,2,3) and (3,4,6), would be the same. However, shifts in the visual matrix, for example, are problematic; a small shift within the image can completely prevent "recognition."

A simple pattern associator ◉ (Fig. 5.3) contains N parallel perceptrons and thus can be used to classify patterns within certain limits. The classes are divided by hyperplanes passing through the origin.

By introducing an additional "bias" neuron... x_0 or also b which has no "sensory input" and always emits the activation 1 (or -1), any planes can be used, i.e., the planes then no longer have to run through the coordinate origin.

Often, analogous to natural neurons, nonlinear activation functions are used to calculate the output activation. This way, partially irrelevant activations can be hidden, or ambiguities can be reduced. The output of a neuron then behaves according to a certain transformation function. This is usually to cause the activity to make a qualitative jump when the input activation exceeds a certain threshold.

For the output of an artificial neuron y_j, we thus obtain:

$$y_j = g\left(\sum_i w_{ij} x_i \right) = g\left(w_j^T x \right)$$

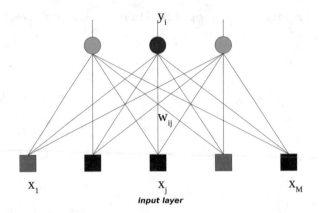

Fig. 5.3 Perceptron as pattern associator

Transfer Functions

The simplest nonlinear transfer function is the threshold function "binary step":

$$g_{step}(x) = \begin{cases} 0 & \text{für } x \leq 0 \\ 1 & \text{für } x > 0 \end{cases}$$

This "jumps" to 1, so to speak, if the input activation is greater than a certain threshold value. θ is. The "ReLU" function ("rectifier linear unit") is similarly simple. It corresponds to a linear function that starts at the coordinate origin and is 0 in the negative range:

$$g_{ReLU}(x) = \max(0, x)$$

To obtain continuous transitions, one uses transfer functions whose graph resembles an S-curve. Especially for backpropagation adaptation in multilayer perceptrons, which we will discuss later, well differentiable activation functions are important. In applications one finds comparatively often the sigmoid function

$$g_{sigmoid}(x) = \frac{1}{1 + e^{-x}}$$

and the tangent hyperbolicus. The range of values of the sigmoid function is between 0 and 1, for the tangent hyperbolicus, it is not only in the positive range, but extends from -1 to $+1$:

$$g_{tanh}(x) = \tanh(x) = \frac{e^{2x} - 1}{e^{2x} + 1}$$

Such nonlinear transfers are also important for the meshes that are supposed to approximate arbitrary functions. Multilayer perceptrons with already two inner layers can approximate any continuous function; if the inner layers contain such sigmoid transfer functions and in the output layer linear transfer functions (Fig. 5.4).

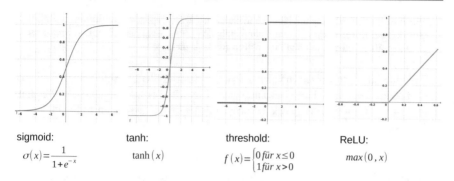

sigmoid: $\sigma(x) = \dfrac{1}{1+e^{-x}}$

tanh: $\tanh(x)$

threshold: $f(x) = \begin{cases} 0 \ f\ddot{u}r \ x \leq 0 \\ 1 \ f\ddot{u}r \ x > 0 \end{cases}$

ReLU: $max(0,x)$

Fig. 5.4 Examples of activation functions (**a**) sigmoid, (**b**) Tanh, (**c**) threshold, (**d**) ReLU

The networks presented so far only generate *local* outputs from the inputs and the weights. In classification problems, among others, it is usually necessary to generate the probabilities taking into account the entire output. For this purpose, we can also apply the well-known SoftMax function, which we already learned about when calculating a probability distribution for probabilistic action selection. However, we have to proceed in two steps. First, the activation of the individual neurons $o_j = w_j^T x$ is generated, and the sum $\sum_k e^{o_k}$ in order to be able to calculate the SoftMax values:

$$y_j = g_j(x) = \frac{e^{o_j}}{\sum_k e^{o_k}} \tag{5.1}$$

Here, j determines a single output neuron at a time and k iterates over all N outputs.

5.1.2 The Adaptability of Artificial Neural Networks

An essential feature of artificial neural networks is their adaptability. The learning process consists of individual learning steps, whereby a new training example is processed in each case. The adaptation usually takes place by adjusting the connection weights. At each learning step, the weights are w_{ij} are adjusted: $w_{ij}^{neu} = w_{ij}^{alt} + \Delta w_{ij}$.

For the calculation Δw_{ij}, there are some basic learning rules. In simple forward linked associators without hidden layers, these are the delta rule and the Hebb rule. For the sake of completeness, I would like to mention the Hebb rule here as well, although it is a learning rule that is applied in principle for "unsupervised" learning scenarios ► (Ch. 1); the delta rule can only be used in the supervised case.

The Hebb Rule
Donald Hebb described in 1949 in *The Organization of Behavior: A Neuropsychological Theory* (Hebb 1949) the following famous connection:

When an axon of cell A is near enough to excite cell B and repeatedly or persistently takes part in firing it, some growth process or metabolic change takes place in one or both cells such that A's efficiency, as one of the cells firing B, is increased.

This means, reduced for the simulation in the computer, that the weight Δw_{ij} of the connection between i and j should strengthen when i and j are active at the same time. For the two-layer network without hidden layers would be the rule for the Hebb update:

$$\Delta w_{ij} = \eta x_i y_j$$

In Hebb learning, the focus of learning is on recurrent pattern contexts. It is assumed that recurring observations are of particular interest.

By adjusting the weights according to the Hebb learning rule, it is achieved that with repeated presentation, the network reacts with increased activity. Looking at the Hebb rule, it is noticeable that the weights either remain unchanged or increase. This results in an unlimited growth of the weights with repeated training. The question of how to deal with this problem in practice arises.

One possibility is to limit the growth to an upper limit w^+ or lower limit w^- and then simply cut off the weights. The obvious disadvantage of this method is that when all the weights have reached one limit or the other, the remaining information content in these weights is very small.

Another option is to normalize the weight vectors, which infer activations of an output y_j after each learning epoch, e.g., with:

$$\mathbf{w}_j \leftarrow \frac{\mathbf{w}_j}{w_j}$$

another would be to set the weight sum to 1, i.e., for all output units j holds $\sum_i w_{ij} = 1$. Such scaling leads to a distributional struggle. Heavily used connections go toward a maximum and unused connections disappear. Furthermore, the normalized weights make the correlation values of representations of different sizes comparable. This is advantageous in competitive learning, where only the winner's representation, i.e., the representation with the "best" pattern correlation, should be adjusted toward the input pattern.

When developing generalized pattern representations, it is important to abstract from the specifics of the individual case. With general "representatives" memory can be saved, and the computation can be accelerated. This is also the origin of an idea to shave weights, which seem to have no meaning for network operations. Typically, these are those that are much smaller than their neighbors. However, this operation requires nonlocal knowledge.

An "overarching perspective" is also taken by another approach, in which negative feedback from the alternative output units enters the learning rule as a decay term (Fyfe 2007):

$$e_i = x_i - \sum_{j=0}^{N-1} w_{ij} y_j \quad mit \; \Delta w_{ij} = \eta e_i y_i$$

This procedure has the advantage that it takes into account statistical properties of the available training data—keyword "principal axis transformation" (PCA)— and furthermore avoids "subsequent" network manipulations.

Such approaches overcome the isolated consideration of individual issues, which is also interesting because it takes into account the constructivist insight mentioned in the introduction that cognitive representations do not represent objective truth but are meant to produce useful distinctions, thus existing only "relative" to the other "representations."

The automatic elicitation of "similarities" is by no means a trivial problem. Often, users of ready-made machine learning frameworks rely on standard constructs without reflecting more closely on their properties. In the case of distinctions, for example, by definition it is important to identify what distinguishes the patterns *from each other*, i.e., the features must be elicited contextually in distinction to the other patterns, whether "o" and "ö" are different has nothing to do with their nature per se, but it depends on whether or not the patterns need to be distinguished. If so, the two punctuation marks make a huge difference; if not, the patterns are virtually the same. This problem cannot be solved in a context-independent manner. A variant of machine learners that specifically address this problem are the so-called support vector machines (SVMs). Here, the separation planes are not determined using all training examples, but only from so-called "support vectors" that lie at the "edges" of the classes to be distinguished. The architecture of classifiers and estimators depends profoundly on the task contexts in which they are to be applied.

Supervised Learning in the Perceptron with the Delta Rule
In the supervised case, to evaluate the output $\hat{y}_j$ at neuron j, we have setpoints y_j. If the output is y_j is too low compared to the setpoint y_j, then for each excitatory input the weight is increased accordingly; if the output is too high, then the weight is reduced. For connections with negative input, the adjustment of the weights takes place in reverse in each case. This results in the following for the update of the connection weights w_{ij}:

$$\Delta w_{ij} = h\left(y_j - \hat{y}_j\right) x_i \tag{5.2}$$

By adjusting the weights with Δw_{ij}, the error over the training set is reduced, i.e., we move downhill on the surface describing the error over the connection weights. Therefore, the delta rule is mathematically a gradient descent procedure. η represents a learning rate that should be chosen suitably small so that no minima are "skipped." The downside of a too small learning rate is that the training takes longer.

After passing a training data set (x, y), the error at an output unit j can be calculated as a function of the connection weights leading to it w_j by:

$$E_j\left(\boldsymbol{w}_j \mid \boldsymbol{x}, \boldsymbol{y}\right) = \frac{1}{2}\left(\hat{y}_j - y_j\right)^2 \tag{5.3}$$

where the estimate $\hat{y}_j$ should be seen as depending on $\boldsymbol{w}$ alone.

We calculate the "descent vector" $\Delta \boldsymbol{w}$ (gradient) by the partial derivatives for each weight. We need to apply the chain rule when deriving the term. This gives us:

$$\Delta \boldsymbol{w}_j = \nabla E_t\left(\boldsymbol{w}_j \mid \boldsymbol{x}, \boldsymbol{y}\right) = \left(y_j - \hat{y}_j\right)\nabla \hat{y}_j \tag{5.4}$$

which, together with $\hat{\boldsymbol{y}} = \boldsymbol{w}^T \boldsymbol{x}$, leads us to the delta rule Gl. 5.2.

If we know the set of training examples, we can also determine the change Δw_{ij} with the summed error of the network over a set of training examples (the so-called batch learning), where E denotes the set of training examples:

$$\Delta w_{ij} = \eta \sum_{t \in E}\left(y_{tj} - \hat{y}_{tj}\right)x_{ti}$$

The general principle followed by the adaptations in the perceptron during supervised learning is always:

$$Change \ = \ learning \ rate \ \cdot (desired \ state - actual \ state) \cdot input$$

A First Perceptron with the "Neuroph" Framework

A good choice for anyone who wants to learn how artificial neural networks work is the free open-source framework "Neuroph." This Java library represents neural networks in an object-oriented way, which makes the structures at all levels very transparent and easily accessible, e.g., also for own "devious" experiments. In the box, an excerpt from the code of the class "Neuron" is shown, in places with [...] code was hidden for clarity, furthermore comments were removed. You can see how classes were explicitly defined for each of the usual basic components of an artificial neural network, such as the neuron but also the layers "Layer," the connections "Connection," or transfer functions "TransferFunction."

Excerpt from the Neuroph Class "Neuron"[2]

```
public class Neuron implements Serializable, Cloneable{
[...]
        protected Layer parentLayer;
        protected List<Connection> inputConnections;
        protected List<Connection> outConnections;
        protected transient double totalInput = 0;
        protected transient double output = 0;
        protected transient double delta = 0;
```

---

[2] Author: Zoran Sevarac; Copyright 2010 Neuroph Project http://neuroph.sourceforge.net Licensed under the Apache License, Version 2.0 (the "License"); http://www.apache.org/licenses/LICENSE-2.0 Further references are in the files of the cited program code.

```
protected InputFunction inputFunction;
protected TransferFunction transferFunction;
private String label;
public Neuron(){
        this.inputFunction = new WeightedSum();
        this.transferFunction = new Step();
        this.inputConnections = new ArrayList<>();
        this.outConnections = new ArrayList<>();
}
[...]
    /**
 * Calculates neuron's output
 */
public void calculate(){
this.totalInput = inputFunction.getOutput(inputConnections);
        this.output = transferFunction.getOutput(totalInput);
}
[...]
}
```

Although such explicit, object-oriented coding has an enormous speed disadvantage compared to frameworks with highly optimized matrix mathematics, also with respect to the available hardware (e.g., the capabilities of current graphics cards), its transparency is very practical, on the one hand, for learners who want to quickly try out their own networks and look at the details, and on the other hand also for the investigation of research questions. The package is small, well documented, easy to use and very flexible. For example, it is easy to access the behavior of individual connections (the "synapses") or neurons. This is helpful, for example, when it comes to experiments to test hypotheses regarding the mode of action of network structures, transfer functions, and so on (Fig. 5.5).

The core of Neuroph consists of the aforementioned Java classes that can map the network structures. To be able to use them yourself, you first need to download the Neuroph framework from the project's download page: http://neuroph.source-forge.net/download.html (19.10.2019). After unzipping the zip file, you will get a folder containing the Neuroph packages, as shown in ● Fig. 5.6.

It's worth taking a closer look at Neuroph's source code to understand the framework and also how neural networks work. The code is designed to be transparent. In the source folder under neuroph/core, you can find the core elements like neuron, layer, and connection. The code snippet from the neuron class shown above shows, for example, how the output of the neuron is generated using the transfer function. In the folder transfer, you will find the implementations of the provided transfer functions. Among them are also those that were introduced at the beginning of the section. The Java code gives exact information about the expected behavior of the individual neurons.

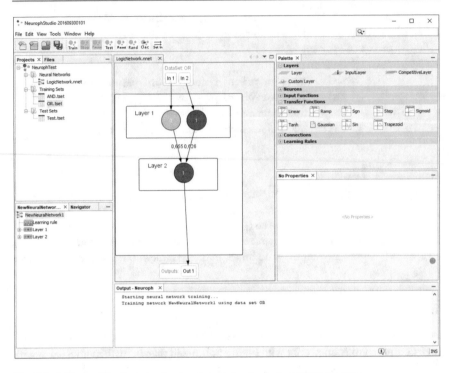

Fig. 5.5 A NeurophStudio trained network mapping the logical "OR" function

Compilation of the Transfer Functions in the Neuroph Framework[3]

```
// TransferFunctionType.STEP ("Binary Step")
public double getOutput(double net) {
        if (net > 0d)
                return yHigh;
        else
                return yLow;
        }}
// TransferFunctionType.GAUSSIANpublic
double getOutput(double totalInput){
        output = Math.exp(-Math.pow(totalInput, 2) / (2*Math.
pow(sigma, 2)));
        return output;
}
// TransferFunctionType.LINEARpublic
double getOutput(double net) {
```

---

[3] Author: Zoran Sevarac; Copyright 2010 Neuroph Project http://neuroph.sourceforge.netLicensed under the Apache License, Version 2.0 (the "License"); http://www.apache.org/licenses/LICENSE-2.0; Further references are in the files of the cited program code.

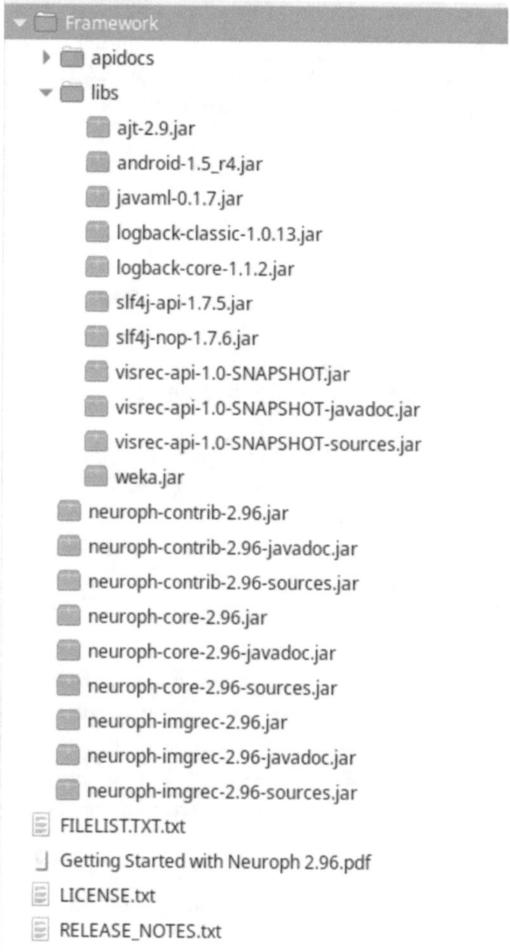

Fig. 5.6 Contents of the
Neuroph package

```
          return slope * net;
}
// TransferFunctionType.RAMP:
public double getOutput(double net) {
          if (net < this.xLow)
                    return this.yLow;
          else if (net > this.xHigh)
                    return this.yHigh;
          else
                    return (double) (slope * net);
}
// TransferFunctionType.RECTIFIED ("ReLU")
public double getOutput(double net) {
```

```
        return Math.max(0, net);
}
// TransferFunctionType.SGNpublic
double getOutput(double net) {
        if (net > 0d)
                return 1d;
        else
                return -1d;
}
// TransferFunctionType.SIGMOID ("Sigmoid")
public double getOutput(double netInput) {
        // conditional logic helps to avoid NaN
        if (netInput > 100) {
                return 1.0;
        }else if (netInput < -100) {
                return 0.0;
        }
        double den = 1 + Math.exp(-this.slope * netInput);     this.
output = (1d / den);
        return this.output;
}
// TransferFunctionType.TANH ("Tangent Hyperbolicus")
final public double getOutput(double input) {
        // conditional logic helps to avoid NaN
        if (Math.abs(input) * slope > 100) {
                return Math.signum(input) * 1.0d;
        }
        double E_x = Math.exp(2.0d * slope * input);
        output = amplitude * ((E_x - 1.0d) / (E_x + 1.0d));
        return output;
}
// TransferFunctionType.TRAPEZOIDpublic
double getOutput(double net) {
        if ((net >= leftHigh) && (net <= rightHigh)) {
                return 1d;
        } else if ((net > leftLow) && (net < leftHigh)) {
                return (net - leftLow) / (leftHigh - leftLow);
        } else if ((net > rightHigh) && (net < rightLow)) {
                return (rightLow - net) / (rightLow - rightHigh);
        }
        return 0d;
}
// The two still existing types TransferFunctionType.LOG and
TransferFunctionType.SIN only use the corresponding Java func-
tions for the
natural logarithm and the sine.
```

In addition, a free GUI tool "NeurophStudio" is also provided on the Neuroph project site, which graphically supports creation, visualization, and storage of neural networks. We will prefer direct Java coding over graphical modeling; the structures are then clearer and more concise, and the behavior of the network can be understood more clearly; in addition we can better couple our networks to the Greenfoot agents. However, to get a first impression and to gain first experiences, a GUI like "NeurophStudio" is quite suitable. You can download the current version of NeurophStudio at http://neuroph.sourceforge.net/download.html (10/19/2019). A suitable JDK is required to start the GUI. It was used JDK 8u221 and the NeurophStudio 2.96 in the presented tests. The framework contains a pdf file "Getting started with Neuroph ..." (cf. ◉ Fig. 5.6). This is a tutorial that allows you to create a simple perceptron that learns the logical functions "and" or "or." Tutorials for this can also be found on video portals such as "YouTube."

Integration in Greenfoot
In order to use external libraries in Greenfoot, they must be copied into the greenfoot/lib/userlib folder. For including Neuroph, it is certainly easiest to copy the jar files from the Framework/libs folder (cf. ◉ Fig. 5.6) into this folder. At least the neuroph-core-xx.jar and the files slf4j-api-1.7.5.jar, logback-core-1.1.1.2.jar, and logback-classic-1.0.13.jar should be included. In the Greenfoot environment, you can check under Tools>Settings>Libraries to see if the copied jar files are listed accordingly.

If everything worked, you should be able to successfully launch the NeurophGreenfoot scenario in the folder "Chapter 5 Estimators for Valuations and Action Selection/ NeuralNetworks_in_Greenfoot."

A window like the one shown in Fig. 5.7 should appear; you may have to create the world manually from the context menu of the LogicFunctions class in the class diagram by selecting the new LogicFunctions() constructor here. When creating the Greenfoot world, if everything worked out with the inclusion of Neuroph, an output will appear in the text console showing how the outputs of the network have been improved by training. In this simple example, we get the new result after just one after a short training period whose duration is in the range of fractions of a second. There should also appear in the "GridWorld" a 4 x 4 grid just as in ◉ Fig. 5.7, which should once again graphically illustrate the behavior of the grid. The "pattern" on the left, consisting of two pixels, is assigned an output of either white (=0) or black (=1).

```
Input: [0.0, 0.0] Output: [0.0]
Input: [1.0, 0.0] Output: [0.0]
Input: [0.0, 1.0] Output: [0.0]
Input: [1.0, 1.0] Output: [0.0]
start time:17-Jun-2021_16-54-48-067_CEST
Training phase...
end time: 17-Jun-2021_16-54-48-068_CESTInput
: [0.0, 0.0] Output:
```

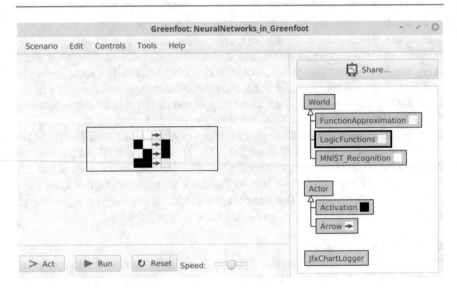

Fig. 5.7 Output of a perceptron trained with the logical function OR in Greenfoot

```
[0.0]
Input: [1.0, 0.0] Output:
[1.0]Input
: [0.0, 1.0] Output:
[1.0]Input
: [1.0, 1.0] Output:
 [1.0]
```

The essential program code for this result can be found in the constructor of the world "LogicFunctions". In this section it is easy to see how Neuroph can be used to create, train, test and store neural networks.

```
public LogicFunctions(){
 super(4, 4, 16); // 4x4 GridWorld, box size 16 pixel
 // generate training data
 trainingSet.add (new DataSetRow (new double[]{1, 0},    new dou-
ble[]{1}));
  trainingSet.add (new DataSetRow (new double[]{0, 1}, new dou-
ble[]{1}));
  trainingSet.add (new DataSetRow (new double[]{0, 0},    new dou-
ble[]{0}));
  trainingSet.add (new DataSetRow (new double[]{1, 1}, new dou-
ble[]{1}));
   // create neural network NeuralNetwork neuralNetwork = new
Perceptron(DIM_input,DIM_output); //MultiLayerPerceptron neural-
Network =
```

```
            new MultiLayerPerceptron(TransferFunctionType.SIGMOID,
2, 3, 1);
 // test untrained testNeuralNetwork(neuralNetwork);
 BackPropagation learningRule = null;
 if (neuralNetwork.getClass()==MultiLayerPerceptron.class){
            learningRule   =   (BackPropagation)neuralNetwork.
getLearningRule();
      learningRule.addListener(this);
      learningRule.setLearningRate(learningRate);
      learningRule.setMaxError(maxError);
            learningRule.setMaxIterations(maxIterations);   }
initJfxLogger();
 System.out.println("start time: "+JfxChartLogger.getTimeStamp());
 System.out.println("Training    phase...    ");    neuralNetwork.
learn(trainingSet); String et = JfxChartLogger.getTimeStamp();
 System.out.println("end    time:    "+et);    jfxLogger.append("end
time; "+et);
 if (learningRule!=null){
      System.out.println("Iterations performed : "+et
                  learningRule.getCurrentIteration()+
                  " Remaining error: "+learningRule.getTotal-
NetworkError()); }
 // test testNeuralNetwork(neuralNetwork);
 // save trained net neuralNetwork.save("kNNs/perceptron.nnet");
 }
```

If you adjust the training set according to the AND function by changing the appropriate entries in the rows where the DataSetRows are created and added, then you should get the expected output for the logical "AND" without any problems.

```
Input: [0.0, 0.0] Output: [0.0]
Input: [1.0, 0.0] Output: [0.0]Input
: [0.0, 1.0] Output:
[0.0]Input
: [1.0, 1.0] Output:
 [0.0]
start time:17-Jun-2021_17-12-45-812_CEST
Training phase...
end time: 17-Jun-2021_17-12-45-813_CESTInput
: [0.0, 0.0] Output: [0.0]
Input: [1.0, 0.0] Output:
[0.0]Input
: [0.0, 1.0] Output:
[0.0]Input
: [1.0, 1.0] Output:
```

[1.0]

The learning process shown was done by a simple (single layer) perceptron. However, if you now try to get the XOR function with a simple perceptron, you will run into difficulties. This is not a software bug but is due to the limitations of the perceptron. Some lines of the code shown above concern multilayer perceptrons and thus already point to the solution. We will clarify this in the section on multilayer perceptrons. Before that, however, we will see for ourselves the already surprisingly great performance of simple perceptrons in recognizing more complex patterns.

The "Hello World" of Supervised Machine Learning: Recognizing Handwritten Digits

A basic problem for machine learning is the recognition of handwritten digits. A famous freely accessible training and test data set is provided by the so-called MNIST database ("Modified National Institute of Standards and Technology database"). It contains about 60,000 handwritten digits in 28x28 pixel format with 256 gray levels.

We can consider each sample pattern as a point in an n-dimensional feature space, whereas with "raw" image data, we rather have to speak of a "perception space," since we have not yet extracted any features, and all input values, in this case the brightness values of the pixels, are equally weighted (Fig. 5.8).

We can consider each digit from this database as a point in a 784-dimensional perceptual space. Pattern classifiers must now approximate interfaces in this perceptual space in such a way that the empirical error over the examples becomes minimal—0 if possible. For practical use, however, it is ultimately not the empirical

Fig. 5.8 Examples from the MNIST dataset https://commons.wikimedia.org/wiki/File:MnistExamples.png (14.12.2017) by Josef Steppan [CC BY-SA 4.0 (https://creativecommons.org/licenses/by-sa/4.0)]

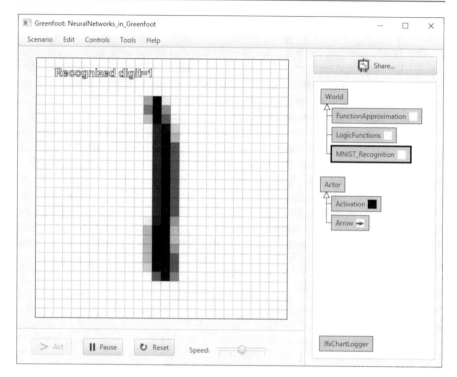

Fig. 5.9 Recognizing handwritten digits with a perceptron in the Greenfoot environment

error that is important, but the so-called structural error, which indicates the probability that an unknown example will be correctly classified.

To launch a perceptron trained with the MNIST dataset, you need to create the world MNIST_Cognition in the NeurophGreenfoot scenario by again selecting the appropriate constructor from the class context menu. An empty 28x28 grid appears in the central GridWorld area. To load the dataset and perceptron, press the Run button. At the end of the training phase, you should see a window like the one in ◉ Fig. 5.9.

The course of the test appears in the text console. This shows how the statistics are produced over the test set. The text output is always in CSV format to allow "copy&paste" into a spreadsheet. The training can now be quite lengthy. At the end of the training, the results of the test run are displayed with the test set.

```
Starting test...
record          num,num,target-output,net-output,correct,wrong,wrong
%,random,random
correct,random wrong,random wrong %
0,1,1,1,0,0.0,1,1,0,0.0
1,5,5,2,0,0.0,6,1,1,50.0
2,1,1,3,0,0.0,6,1,2,66.66666666666667
```

```
3,7,7,4,0,0.0,2,1,3,75.0
```

The simple perceptron shows a structural error of about 20% with an unknown test set, the empirical error over the training set is about 14.7%. For a purely random selecting system, an error of about 90% is to be expected with ten classes. For comparison, a purely random selection is still performed and from CSV column 8 "..., random ,..." is also shown in the statistics.

We recall that such a perceptron has no inner layers "hidden layers." The 784 input neurons are directly connected to the output layer. On the MNIST database website, linear classifiers are mentioned, which with appropriate settings achieve an error rate of less than 10%. You can also train a network yourself in the prepared scenario. To do this, in the method "public void started()," which is called when the Greenfoot "Run button" is pressed, you only have to comment out the line with the function "loadANN" and comment in the line with the call to the function "train." The method should then look something like what is shown in the code example below.

```
@Overridepublic
void started() {
System.out.println("greenfoot started "
        +JfxChartLogger.getTimeStamp()+")");
if (jfxLogger==null) {
        initJfxLogger();
}
if (!netIsTrained){
            //loadANN("ANNs/mnist_perceptron.nnet",  "data/mnist_
test200.csv");
            train("data/mnist_training1000.csv",  "data/mnist_
test200.csv");
}
}
```

During the creation of the world, you get outputs in the text console, which represent the learning progress. Files are prepared that contain somewhat "slimmed down" training and test sets. The smaller training set "mnist_training1000.csv" contains the first 1000 examples from the MNIST set and the file "mnist_test200.csv," the following 200 records for the test. You will also find more extensive files in the Data folder, but the training process will of course take correspondingly longer.

The preset network was trained with 40,000 examples for 2000 epochs, using a learning rate of $\eta=0.001$. However, the training run took several hours on a standard PC, whereby the inefficient structure of the Neuroph library in terms of resource consumption also has a negative impact. In addition, the calculations can actually be parallelized very well, but neither the possibilities of a multicore CPU nor a GPU are used for parallel computing.

If you wish, you can trace the training by providing the algorithm with the appropriate files. You can find them in the subfolder "data," which is located in the folder of the scenario and where the logging files are also saved. The learning rate η has a relatively large impact on the error rate, which may be due to the fact that very fine nuances are crucial for the network's decision, especially at the end of the training. If the adjustment is too "coarse," the necessary fine adjustments cannot be made sufficiently, like an old radio receiver with a frequency control that is too coarse: The learning process adjusts back and forth, but clear reception never occurs.

You may be able to increase the efficiency of the mesh by improving parameters or alternative mesh properties. By grabbing the CSV-formatted text outputs with copy&paste, you can also produce outputs like those in ● in any spreadsheet (Fig. 5.10).

Limits of the Perceptron
The performance of the perceptron in pattern recognition is quite impressive. However, the limits become clear already by simple examples. Maybe some readers already know that it is not possible for a perceptron to learn the simple XOR function.

If you create the world "LogicFunctions" again in the "NeuralNetworks_in_ Greenfoot" scenario, then you can try this by adjusting the training set in the constructor according to the XOR function. You will find that the algorithm fails to bring the error below the desired level. What is the reason for this? The problem becomes clear if you plot the two-dimensional domain of definition of the functions in a coordinate system (Fig. 5.11).

Perceptrons discriminate on the basis of hyperplanes that have one dimension less than the input space. For the two-dimensional case, these can be represented by a straight line. However, the training examples for the function XOR are not separable by means of a linear function. To solve this problem with an artificial neural network, it is necessary to insert additional "hidden layers" between the input and output layers. However, this also requires an extension of the learning rules.

Fig. 5.10 Learning curve of the perceptron with the MNIST training set

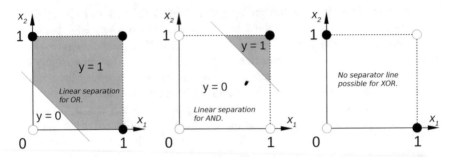

Fig. 5.11   Linear separability of the elementary logical functions

5.1.3   Backpropagation Learning

With the presented artificial "neural" elements and the open library, a variety of alternative network topologies can be generated. In this area, exciting experiments can also be carried out by "laymen" and real basic research can also be conducted. However, freely connected networks quickly exhibit an astonishing complexity that is sometimes difficult to control.

A feedforward layered design facilitates the simulation of parallel processing and maintains mathematical controllability; moreover, well-studied methods for supervised adaptation exist. Therefore, multilayer "feedforward" perceptrons (MLPs) are mostly used in practice. In this field, one has meanwhile strongly detached oneself from the biological models and investigates these networks with the means of mathematical statistics.

For multilayer perceptrons, there exists an extension of the delta rule, the previously mentioned backpropagation method ◉ (Fig. 5.12). The "backpropagation" algorithm is applied to networks with hidden intermediate layers, i.e., networks with neurons without direct connections to the input or output layers. Like the delta rule, the backpropagation method belongs to supervised learning, according to which adaptation takes place according to the difference between the input and the realized output. Learning practically takes place in three stages.

Forward Pass

The input neurons receive stimuli with which a preliminary output is computed.

Fault Determination

At each output unit, the difference between the desired and actual output is calculated. If there is an overrun of the error tolerance, the backward pass is executed in the third step.

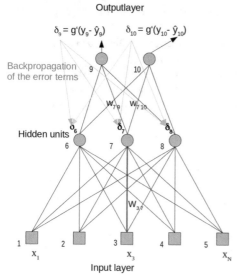

Outputlayer

$\delta_9 = g'(y_9 - \hat{y}_9)$ $\delta_{10} = g'(y_{10} - \hat{y}_{10})$

Backpropagation of the error terms   9   10

$w_{7\,9}$ $w_{7\,10}$

δ_7 δ_8

Hidden units   6   7   8

$w_{3\,7}$

x_1 2 x_3 4   5 x_N

Input layer

1. Foreward-Pass: The network generates "preliminary" outputs (y_9 and y_{10}).

2. 'Error' calculation: The error terms are formed from the differences between target and actual values at the output layer.
$$\delta_9 = g'(y_9 - \hat{y}_9)$$
$$\delta_{10} = g'(y_{10} - \hat{y}_{10})$$

3. Backward-Pass: 'Back propagation' of the error. Two example calculations on the unit 7:
$$\Delta w_{7\,9} = \eta\, y_7\, \delta_9$$
$$w_{7\,9}^{(new)} \leftarrow w_{7\,9}^{(old)} + \Delta w_{7\,9}$$

$$\delta_7 = \eta\ g'(\delta_9 w_{7\,9} + \delta_{10} w_{7\,10})$$
$$\Delta w_{37} = \eta\, x_3\, \delta_7$$
$$w_{37}^{(new)} \leftarrow w_{37}^{(old)} + \Delta w_{37}$$

Fig. 5.12 Backpropagation adjustment

Backward Pass

The error terms are propagated backward to the input layer. At each step, these decrease. Subsequently, the update of the weights in the hidden layers is carried out with these error terms.

The backpropagated error δ_j on a hidden unit j is the weighted sum of the errors δ_l of all units l receiving an input from unit j multiplied by the derivative of the activation function g.

$$d_j = \begin{cases} y_j - y_{\wedge j} & \text{if } j \text{ is an output unit} \\ g'\left(\sum_k d_k w_{jk}\right) & \text{if } j \text{ is a hidden unit} \end{cases}$$

The parameter y_i is the output of the corresponding antecedent neuron. In the case of linear activation functions, the derivative g' is a constant which can be formally included in the learning rate η. Finally, the adjustment of the weights is done with $\Delta w_{ij} = \eta y_i \delta_j$.

You can see the backpropagation algorithm in action, if you replace the line:

```
NeuralNetwork neuralNetwork = new Perceptron(2, 1);
```

by this:

```
MultiLayerPerceptron neuralNetwork = new
    MultiLayerPerceptron(TransferFunctionType.SIGMOID, 2, 3, 1);
```

You achieve this favorably by commenting in and out the appropriate prepared lines.

This creates an MLP with sigmoid transfer functions, which has two neurons in the input layer, three hidden units and a single output unit.

With a learning rate of $\eta = 0.05$ and a maximum error of 0.001, the network reaches the desired accuracy after 36,412 iterations. One can clearly see the output changes of the network in the test outputs before and after the training phase.

```
Input: [0.0, 0.0] Output: [0.35445948948784073]
Input: [1.0, 0.0] Output: [0.3409904933860178]
Input: [0.0, 1.0] Output: [0.3438499793399638]
Input: [1.0, 1.0] Output: [0.3316392455492841]
start time:17-Jun-2021_18-49-30-474_CEST
Training phase...
iteration 5000: 0.12568209975580955
iteration 10000: 0.12563959681019288
iteration 15000: 0.12531262329534443
iteration 20000: 0.08914230502123272
iteration 25000: 0.0060983880224569345
iteration 30000: 0.002156014247987953
iteration 35000: 0.0012491561662580403
end time: 17-Jun-2021_18-49-30-644_CEST
Iterations performed :37834 Remaining error:9.9994637943421E-4
Input: [0.0, 0.0] Output: [0.04635849253970898]
Input: [1.0, 0.0] Output: [0.9568814261136893]
Input: [0.0, 1.0] Output: [0.9568927922736467]
Input: [1.0, 1.0] Output: [0.04615368654491579]
```

Deep Learning

However, the recognition of patterns can often not be solved satisfactorily with "simple" MLPs. As mentioned above, simple transformations such as shifting or scaling may prevent recognition completely. Before using a network, one should consider to which kinds of changes the classifier should be "insensitive." So-called "deep networks" have at least two hidden layers and can extract local features in patterns and combine them hierarchically. These "deep" neuronal networks belong to the technically most complex models of artificial neuronal networks, such as the "Neocognitron" already published in 1980 (Kunihiko Fukushima 1980). With the help of a large number of hierarchically combined networks, an attempt is made here to establish insensitivity to transformations such as shifting and scaling. As a result, these networks can recognize hierarchical, structured spatial formations, i.e., patterns where simple spatial elements form more complex structures, which in turn are elements of even more complex patterns—invariant to displacement and thus also invariant to multiple deformations. More modern versions, the so-called convolutional neural networks, solve the problem with the help of the so-called convolutional layers, which reduces the resource consumption enormously. At the same time, the available hardware (e.g., graphics cards, main memory, multicore CPUs) has been developed enormously, especially for the necessary calculations, so that the sometimes spectacular results could be achieved. With these capabilities, such networks can be used for many tasks in which the processing of complex patterns plays a role, e.g., in the recognition of handwriting, the evaluation of audio signals, or in the identification of objects in photos or camera images, to name just a few common applications.

Using a large number of hierarchically combined meshes, an attempt is made here to generate insensitivity to transformations such as displacement and scaling. This allows these networks to recognize hierarchically structured spatially extended patterns—i.e., patterns in which simple spatial elements form more complex structures, which in turn are elements of even more complex patterns—invariant to displacements and thus also invariant to multiple deformations. Some current versions, of ANNs called "Convolutional Neural Networks" (CNNs), solve the problem using so-called convolutional layers, which reduces resource consumption enormously. Since at the same time the available hardware (e.g., graphics cards, main memory, multi-core CPUs) for the necessary calculations has been enormously advanced, the sometimes spectacular results could be achieved.

With the aforementioned capabilities, such networks can be used for many tasks involving the processing of complex inputs, such as handwriting recognition, audio signal analysis, or the identification of objects in photographs or camera images, to name just a few common applications.

Our goal in the following is to use MLPs in our reinforcement learning algorithms as estimators for state evaluations or for action selection. Before we do so, we will take a closer look at their ability to approximate arbitrary functions.

5.1.4 Regression with Multilayer Perceptrons

MLPs with sigmoid transfer functions with two hidden layers can already approximate all continuous functions (Frochte 2019). This capability can also be used for nonlinear regression analysis. Regression analysis is concerned with fitting a curve as optimally as possible through a given point cloud. The goal here is to minimize the squared error over the data set. Often a certain type of function is used as a mathematical model, e.g., a linear, a power or an exponential function. In regression analysis, the parameters of the model are adjusted to minimize the squared errors. In regression with multilayer perceptrons, such a basic model is not necessary. This can be used in use cases with complex data where it is not easy to find a suitable model, such as financial market information.

In the scenario "neural networks in Greenfoot," an MLP is stored that can be trained with the course of a stock index (e.g., the annual course of the DAX in the example). The goal of the training is to minimize the corresponding prediction error of the network. In the subfolder "data," there is a CSV file that contains the daily closing prices of the DAX from 11/1/2018 to 10/31/2019. Since this price history is one-dimensional, the default multilayer perceptron has only one input and one output neuron. However, the hidden layer consists of 20 neurons with sigmoid transfer function.

Create the world "FunctionApproximation" for this. The Greenfoot grid is not actually used this time. The important outputs are done in the console and after the training is finished in the "jfxLogger" window. The training starts when the Run button is pressed.

You are welcome to try out other configurations, e.g., with several hidden layers. The default learning rate is $\eta = 0.01$. The example does not claim to be optimal. It is quite generally associated with some luck and intuition to find an efficient network configuration for the given task. This also becomes clear in the examples presented in the following.

For example, the learning rate η has a large impact on the training progression. For diverse learning rates η, a training progression like the one shown by the red curve in ◉ Fig. 5.13 (learning rates $\eta = 0.001$). This was usually tested with an upper limit of 100,000 episodes in order to be able to sample different values in a feasible computation time. The "red progression" into the sink at an error of about 0.0037 also applied to some test runs with the rate $\eta = 0.01$, which was later found to be quite favorable. Randomization of the connection weights results in different training progressions. In one run, the network suddenly found itself out of the local sink after about 50,000 episodes and descended to an error of about 0.0016 (yellow curve). The training shown ran for only 100,000 episodes, so the curve is horizontal after 100,000 episodes.

With a finer learning rate $\eta = 0.001$, the network does not find its way out of the 0.0037 dip even 1,000,000 episodes (red curve). With another pass with $\eta = 0.01$ and an upper limit of 1,000,000 episodes, the network descended to an error of about 0.0012 (green curve). In this case, the exit from the dip happened rather suddenly only after about 245,000 episodes. In the figure ◉ (Fig. 5.14) one can see that

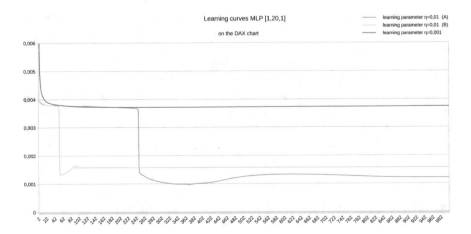

Fig. 5.13 Learning curves in the regression of the DAX annual trend

A German Share Index (DAX) - Annual performance and regression curves of a MLP [1,20,1]

Fig. 5.14 Different MLP regression curves in the DAX annual course

the "better" curves (yellow and green) differ significantly, although the same learning rate was used. The aspect of generalization capability, or "overfitting," must also be taken into account. Sometimes an exact reproduction of the original is not desired, since the regression also serves to abstract from random fluctuations. A too exact mapping reflects random fluctuations that have no meaning for the application, and the extrapolation capabilities of the network may then also be reduced.

MLPs can of course also be used to estimate higher-dimensional functions. For this purpose, only the number of input neurons in the input layer has to be increased.

Advanced Java Technology
In principle, this training, which falls into the area of supervised learning, does not really need the Greenfoot environment. Much faster is the training with the use of a professional framework, such as the amazon open-source "Deep Java Library" DJL (https://djl.ai/ ;30.6.2021). DJL also implements the java principle "write once and run anywhere," which means with DJL you can not only choose on which of the popular engines, like TensorFlow, PyTorch, mxnet, onnx, and more, the java code should be executed, it is also possible to use different engines at the same time in a java application. A practical possibility also for professional developers to compare the performance in the respective use cases and to choose the optimum. Feel free to deepen your knowledge in the area of deep learning! You can also find exciting tutorials on the site (https://docs.djl.ai/jupyter/tutorial/index.html ;30.6.2021).

In any case, the capabilities and technical complexity of the approximators can be significantly increased. For didactic and content reasons, we now turn back to reinforcement learning, where we will use the ability of MLPs to approximate arbitrary functions to parametrically estimate our agent functions that we have implemented tabularly so far, such as the state evaluation function or the policy of the agent program.

Even with our "simple" tools, we can already obtain some interesting results and illustrate how such approximators are used in reinforcement learning algorithms. In the following, we will approximate the state evaluation function, i.e., the function that assigns an evaluation to the sensory inputs, in the context of an extended Q- or Sarsa-learning using multilayer perceptrons.

5.2   State Evaluation with Generalizing Approximations

So far, we have stored the scores of the state-action pairs in a table, which is the simplest way to store an assignment. For very large state-action spaces, such a table is not practical, since we can no longer store values for every possible state due to the limited availability of time and memory. Such tables are also implausible with respect to "real environments" or the biological originals; imagine, for example, the situation with optical or other sensors that provide tens of thousands of continuous values for each state.

In such state spaces, we can first make do by treating the estimation of the Q-values as a regression problem and using the valuation func-tions $\hat{Q}(s,a,\theta) \approx Q^\pi(s,a)$, resp. $\hat{V}(s,\theta) \approx V^\pi(s)$ with supervised learning meth-ods, e.g., artificial neural networks, where the parameter vector θ, e.g., can stand for the connection weights of a kNN. However, other types of numerical function approximations can also be used.

In principle, the error over the states results from the squared error over the state space:

$$E(\theta) = \sum_{s \in S}\left[V^\pi(s) - \hat{V}(s,\theta)\right]^2$$

The state space contains many states that are of little interest to us. Since it makes little sense to train the estimator on "unimportant" states and possibly even neglect the evaluation of "interesting" states in order to do so, in Sutton and Barto (2018) the formal prediction quality with a kind of "interest distribution," which is weighted $\mu(s)$ which weights which states are important with respect to the prediction accuracy.

$$\overline{VE}(\theta) = \sum_{s \in S}\mu(s)\left[V^\pi(s) - \hat{V}(s,\theta)\right]^2$$

Often $\mu(s)$ is identified with the relative length of stay in a state or with the rela-tive number of visits. Unfortunately, we also cannot assume that the adjustments of the estimator reproduce the nonlinear structure of the value function; this means that by improving the estimation of one state we may get a deterioration elsewhere. This is one of the central technical problems in the so-called deep reinforcement learn-ing, i.e., reinforcement learning with deep neural networks. Numerous proposals have been published on this recently. However, we will limit ourselves to the algorithms described and implemented in tabular form before, in order to build on what is known and to maintain a certain consistency in terms of content.

Learning the Estimation Function from the TD Error

Usually the number of weights (the dimensionality of θ) is much smaller than the number of world states. Moreover, with the help of such a regressor, we can evaluate (s,a)-pairs even if they have not occurred so far.

In temporal difference learning methods such as Sarsa(0), we try to find the difference of $Q(s_t, a_t)$ and $r_{t+1} + \gamma Q(s_{t+1}, a_{t+1})$ to be reduced. With this task, we can also train a supervised learner, where the input is $x_n = (s_t, a_t)$ and the required output $y_n = r_{t+1} + \gamma Q(s_{t+1}, a_{t+1})$ is assigned to the input. The squared TD error at time t can be calculated with:

$$E_t^V(\theta) = \left[r_{t+1} + \gamma \hat{V}\left(s_{t+1}, \theta\right) - \hat{V}(s_t, \theta) \right]^2 \tag{5.5}$$

respectively

$$E_t^Q(\theta) = \left[r_{t+1} + \gamma \hat{Q}\left(s_{t+1}, a_{t+1}, \theta\right) - \hat{Q}(s_t, a_t, \theta) \right]^2 \tag{5.6}$$

can be described. In principle, we could use all possible regression methods to learn the valuation function. However, there are also some challenges in constructing the supervised learner for the valuation functions $\hat{Q}(s,a,\theta)$, respectively $\hat{V}(s,\theta)$.

"Similar" state-action pairs should also have similar valuations, which requires estimators that also work out the corresponding characteristic features in the learning process. Moreover, it is the case that the change of one parameter influences the estimation value of very many states.

In the tabular case, we could assume that the learned scoring function corresponded to the exact values. To simplify, the values for each state were learned independently of the others. An update at one state did not affect any other state. However, if we use an estimator, an update at one state simultaneously affects many other states. This requires, in part, a lot of "finesse" in the selection and setting of the learner.

Moving from simple Q-learning to "deep Q-learning" sometimes turns out to be more problematic than expected. In general, learners with small learning rates or those capable of building local models, e.g., "kernel-based" approximators, such as RBF estimators, are recommended (Alpaydin 2019, Ch.18.6). Other technical measures are also recommended to stabilize the learning process, e.g., "experience-replay." In this case, training data are first stored in bundles and finally learned in packages ("batch learning"), whereby the training data are randomly mixed.

Since the breakthrough of deep learning, the corresponding multilayer perceptrons or convolutional neural nets with their ability to perform nonlinear function approximations are increasingly preferred for large state and action spaces. In particular, they show their strength when processing visual data, up to "raw pixels" as input. However, these networks are not necessarily the best choice when it comes to appropriate estimations.

While generalizability makes learning with such estimators potentially more powerful, it also tends to be more difficult to manage and understand. The question of how we extend machine learning systems, especially in "deep learning," to make it transparent to us what their decisions are based on—keyword "Explainable AI") (Been and Pavlus 2019)—is becoming increasingly important.

If we now use a gradient descent method, as with ANNs, then we update the parameter vector with:

$$\Delta q = h\left[r_{t+1} + g\hat{V}\left(s_{t+1},q\right) - \hat{V}\left(s_t,q\right) \right] \nabla_{q_t} \hat{V}\left(s_t,q\right) \qquad \text{(Gl. 5.7)}$$

or with

$$\Delta q = h\left[r_{t+1} + g\hat{Q}\left(s_{t+1},a_{t+1},q\right) - \hat{Q}\left(s_t,a_t,q\right) \right] \nabla_{q_t} \hat{Q}\left(s_t,a_t,q\right) \qquad \text{(Gl. 5.8)}$$

for individual actions. The equations can be derived from Gl. 5.5 resp. Gl. 5.6, respectively.

One algorithm that uses these updates is the "semi-gradient Sarsa" (Sutton and Barto 2018) Fig. 5.15.

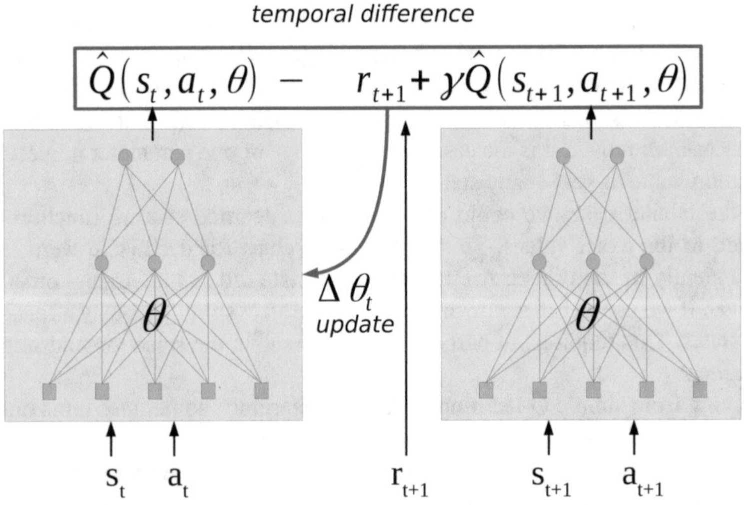

Fig. 5.15 Learning a Q-value estimation with the semi-gradient Sarsa

Semi-Gradient Sarsa-Learning Algorithm According to R. Sutton and A. Barto

```
1       initialize θ ∈ ℝ^d arbitrarily (e.g. θ=0)

2     Loop for each episode

3             initialize s,a with initial state and action

4             Repeat

5                   take action a, observe r and s'

6                   if s' is terminal:
```
$$\theta \leftarrow \theta + \eta[r - \hat{Q}(s,a,\theta)]\nabla\hat{Q}(s,a,\theta)$$
```
                    Start next episode

8                   Choose a' in respect to Q̂(s', ·, θ) and the action selection
                    strategy of the policy (e.g., ε-greedy or SoftMax).
```
$$9 \qquad \theta \leftarrow \theta + \eta[r + \gamma\hat{Q}(s',a',\theta) - \hat{Q}(s,a,\theta)]\nabla\hat{Q}(s,a,\theta)$$
$$10 \qquad s \leftarrow s', a \leftarrow a'$$
```
11            While s is not terminal
```

In the procedure presented, no true gradient descent takes place, which is why it is also referred to as "semi-gradient." One problem in this procedure is, for example, that while the effects on the estimation that arise from changing the weights are taken into account, any effects on the objectives pursued are ignored.

Although these "bootstrapping" methods ("bootstrapping"-meaning "pulling yourself out of the swamp by your own hair") do not converge as robustly as true gradient methods, they do converge reliably in important cases. Moreover, they usually allow much faster learning and allow continuous, "on-line" learning without waiting for an episode to end (Sutton and Barto 2018).

It is also important to consider the properties of the estimator $\hat{Q}(s,a,\theta)$ to be taken into account. As already mentioned, such a one-step update places special demands on the estimator and the design of the learning, since many "experiences" may arrive that are not very relevant or contain little or even counterproductive information. This can also worsen a given estimation, since such "observations" may then be inadmissibly generalized.

"Semigradient Sarsa" Update in Java

```java
protected void update( State s, int a, double reward, State s_
new, int
a_new, boolean episodeEnd ){
 double observation = 0.0;
 if (episodeEnd) {
      observation = reward;
 }else{
      // Gets approx. Q from MLPs:
      observation = reward + (GAMMA * getQ(s_new, a_new));
```

```
   }
  setQ_toMLP(s, a, observation);
 }
 protected double getQ(State s, int a){
  double[] input = s.holeFeatureVector();
  neuralNetwork[a].setInput(input);
  neuralNetwork[a].calculate();
  double[] output = neuralNetwork[a].getOutput();
  return State.originalValue01(output[0], minimumQ, maximumQ);
 }
 protected void setQ_toMLP(State s, int a, double observation){
  double[] input = s.getFeatureVector();
  double[] target_output = new double[DIM_output];
  target_output[0]= State.range01(observation,minimumQ,maximumQ);
  DataSet trainingSet = new DataSet(DIM_input, DIM_output);
  trainingSet.add(new DataSetRow(input, target_output));
  neuralNetwork[a].learn(trainingSet);
 }
```

In the example, the neural "memory" of the agent was not only implemented with a multilayer perceptron, but a separate MLP was provided for each possible action. The MLPs each provide the value of an action in a given state. This procedure works quite well in tests. Intuitively, the finding can also be explained by the fact that pattern components that make the alternatives interdependent are hardly productive. Such an action selection is after all a decision with either-or. For living beings it is not possible to flee or fight or to bite into a fruit and reject it at the same time. Maybe a competitive arrangement would make sense, e.g., with a pairwise comparison of each possible alternative. Readers are welcome to experiment with special constructions.

Some success can even be achieved in this scenario by feeding the sensory values, altitude, speed, and tank content, directly into the network, but the process becomes much more stable if we construct a feature vector similar to ▶ section "Monte-Carlo Policy Gradient (REINFORCE)". This can be done in several ways. In the simplest variant, the continuous space is divided into equal intervals and a 1 is set in each case if the value lies within the corresponding limits and 0 otherwise. Sometimes, however, this does not achieve the necessary resolution to generate a stable learning process leading to the optimum.

Another option is to use radial basis functions (RBFs). The maximum of these functions is located at the "prototypical" centroids of the feature and then falls off according to the "kernel" function used, with increasing distance from this center. Often the Gaussian function is used here, which provides an optimal description for the distribution of values that deviate from a given center purely by chance, such as occurs with measurement errors. A typical RBF feature provides a bell-shaped signal strength, which depends on the distance between the value s_i and the center state c_i of the feature:

$$x_i = RBF\left(s_i\right) = e^{-\frac{\left(s_i - c_i\right)^2}{2\sigma^2}}$$

The characteristics can of course also be defined multimodally, so that in each case instead of the difference $s_i - c_i$, a distance measure of whatever kind $d(s, ci)$ in the "perception space" is taken as a basis for the generation of the feature signal.

In the function available in the Java example, in the value ranges of $[0; 1]$, a certain number of centroids (6) are evenly distributed (Fig. 5.16).

Although RBFs produce continuous approximations that generate continuous and intuitively plausible values with respect to the presence or absence of a feature, moreover they are also well differentiable, in practice they have been found to require a relatively high computational effort and can also reduce performance, especially for more than two state dimensions (Sutton and Barto 2018).

In Sutton and Barto (2018) binary features are favored, where the input space is divided into "tiles." Each tile represents a component of the feature vector. A "one" is now written for the tile in which the measured value is located, otherwise "zero." To increase the quality of this feature vector, several layers of tiles are stacked and each is shifted slightly. This increases the number of "ones" in the feature vector to the number of corresponding layers. It is recommended that the shift is not linear, but according to an "offset" vector, e.g., $(1\ 3)^T$ (in 2D) or $(1\ 3\ 5)^T$ (in 3D) etc (Fig. 5.17).

We test our algorithms in the "Lunar Lander" scenario, which is usually included with the Greenfoot package as an introductory example ◉ (Fig. 5.18). This is a simple, playful simulation of a lunar landing. The task is to decelerate a lunar module falling onto the lunar surface with targeted rocket impulses in such a way that the landing speed is minimal. The sensors provide the parameters altitude, speed, and tank content. The reward depends on the landing speed. If the landing speed falls below the maximum load limit (MAX_LANDING_SPEED), the landing is successful; otherwise, the lander is destroyed.

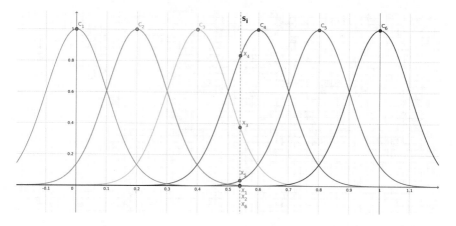

Fig. 5.16 Generation of a feature vector **x** from **s** using RBF features

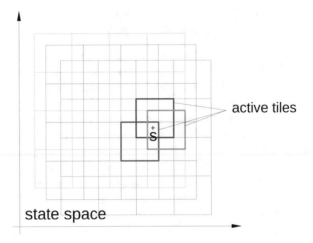

Fig. 5.17 Representation of a state s by multiple overlapping and shifted subdivisions. Counting from the top left, tile 12 is active in the first level (orange), tile 11 in the second level (red), and tile 7 in the third level (blue). "Active" tiles receive the value 1, the others 0

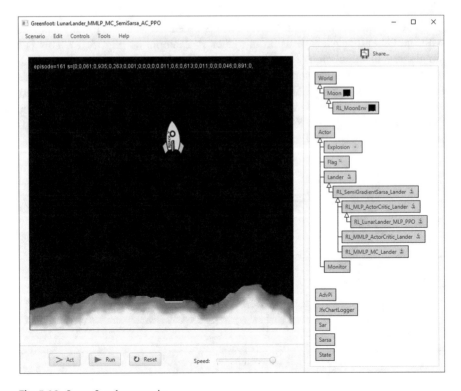

Fig. 5.18 Lunar Lander scenario

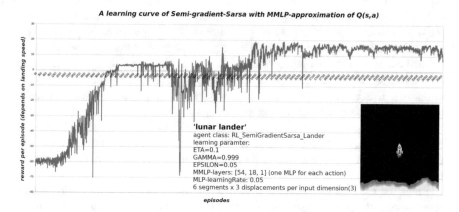

Fig. 5.19 A learning curve for Lunar Lander with semi-gradient Sarsa ($\gamma = 0.999$; $\eta = 0.1$; $\varepsilon = 0.05$). One MLP per action option was used. MLP layers: [55,18,1]. The range of sensor values was divided into six binary features and shifted three times

In the Lunar Lander example, the work was not multimodal. Each individual sensory area was divided into six intervals each. If the "measured value" is within the respective section, the feature vector receives a 1 at the respective position. With 3 sensors, we would thus receive 18 features that always contain three "ones" and 15 "zeros." For the illustrated experiment, the vector was extended by three shifts according to the above principle, resulting in a binary feature vector of size 54 (Fig. 5.19).

Regarding the reward, one might think that it would be appropriate to give a "bonus" if the landing was successful. In tests, however, it had been found that it is more beneficial for the training of MLPs if the reward function is steady, i.e., does not make "jumps." The design of the reward function can have a great influence on the success of the learning process and can also take some time of "trial and error."

Reward Function in Lunar Lander

```
private double checkReward(){
        double reward = 0;
        if (isLanding()||isExploding()){
            reward= MAX_LANDING_SPEED-((double)speed);
        }
        // bounds
        if (reward>maximumQ) reward = maximumQ;
        if (reward<minimumQ) reward = minimumQ;
        return reward;
}
```

The state-action-value function usually does not show a continuous behavior. Small deviations in the input can make a big difference in the evaluation, for example, when it comes to the presence or absence of an important feature. It is also possible that sensory values have no relevance at all in large value ranges, but from a critical point on they decide everything, e.g., in the case of the tank content of the Lunar Lander. At first this sensor value has no meaning for the design of the trajectory. However, if the fuel level falls below a critical level, the result is "suddenly" crash. In a distributed learner, something like this can cause some of the previously learned evaluation quality to be diminished.

To stabilize the learning process with kNNs, the measures mentioned, such as "experience-replay," are recommended. The random order in the processing of the stored "experiences" is intended to prevent certain recurring patterns from leading to undesired artifacts in the training of the network.

Since we have an episodic scenario, we can also use the Monte Carlo update. Here we can take advantage of the fact that our final reward is a secured observation.

In the Java implementation of the update, it can be seen how the recorded episode is mined from the back, the scaled and accumulated "utility" G is permanently updated. The "experience" evaluated in this way is inserted at a random position in the training dataset to generate a random sequence of training data.

Implementation of a Gradient Monte Carlo Update for a State Value Estimator in 'LunarLander' Scenario

```
protected void update(LinkedList <Sar> episode){
 trainingDataReset();
 double G=0;
 while (!episode.isEmpty()){
        Sar e = episode.removeLast();
        State s_e = e.s;
        G=GAMMA*G+e.reward;
        double[] target_output = new double[DIM_output];
        target_output[0]=State.range01(G,minimumQ,maximumQ);
        // insert at random position:
        int size = trainingSet[e.action].size()+1; trainingSet[e.
action].add( random.nextInt(size),
                        new  DataSetRow(s_e.getFeatureVector(),
target_output));
 }
 adjustMLPs();
}
```

"Example Mountain Car"

A common standard scenario is the "Mountain Car Task" by Sutton and Barto (2018). This involves getting a car over a steep hill with insufficient engine power. The difficulty is that gravity is stronger than the car's engine can generate in terms of acceleration, and the car can't get over the steep slope even at full throttle. The

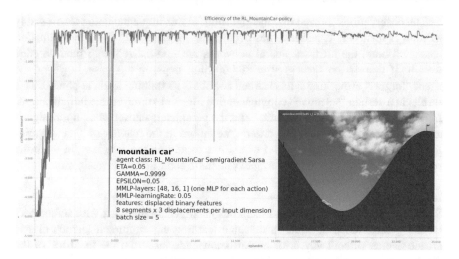

Fig. 5.20 The "Hill climb" scenario in Greenfoot

only solution is to first drive backward away from the target and use the opposite slope on the left to gather enough potential energy to then, under full throttle and using the car's inertia get over the hill. In this example, the catch is that the agent is in a situation where things have to get worse, so to speak, before they can get better. Many control methods have great difficulty accomplishing tasks of this nature without resorting to human assistance (Fig. 5.20).

The reward in this problem is −1 at each time step until the car passes its target position at the top of the mountain, which ends the episode. There are three possible actions: full throttle forward, full throttle reverse, and simple disengaged roll. The car moves in the scenario according to simplified physics. A state is defined by the position x_t and the speed $\dot{x}_t$. The position x_t and the velocity $\dot{x}_t$ follow

$x_{t+1} \doteq x_t + \dot{x}_{t+1}$ (whereby x_{t+1} is limited to −1, 2 $\leq x_{t+1} \leq$ 0, 5 is limited) and $\dot{x}_{t+1} \doteq \dot{x}_t + 0,001A_t - 0,0025\cos(3x_t)$ (where   for   the   speed $\dot{x}_{t+1}$ applies: $-0,07 \leq \dot{x}_{t+1} \leq 0,07$).

This corresponds to a "hilly landscape," which follows the curve of $y = \sin(3x)$ in the mentioned interval limits. Furthermore, in the scenario the speed is set to 0 when the left limit is reached. At the interval boundary on the right, the destination is reached and the episode ends. Each episode starts at a randomly chosen position x_0 with $x_0 \in [0, 6; 0, 4[$ and a starting speed $\dot{x}_0 = 0$.

Surprisingly good results were obtained with an arrangement in which the networks with two input neurons each were directly given the values for position and velocity. This input layer was then followed by 16 neurons in the first hidden layer and 4 in the second. The idea behind the rather large number of neurons in the first hidden layer was to allow local features to emerge that favor one action or the other. In the output layer, there is only one neuron, each of which outputs the estimated Q value.

The observations showed that the algorithm in this implementation learns stably as soon as a success experience has occurred at least once. In the implementation presented here, the artificial neural networks are reinitialized with random edge weights if there is no success after 10,000 time steps. In this case, however, the agent "forgets" everything it has learned and starts its training again at point 0. This results in a certain "selective" component that filters out unfavorable initializations. Intuitively, then, the implementation and the parameters chosen were not designed and tried according to scientific criteria, i.e., more in the manner of "manual reinforcement learning." Readers who would like to rely on the data, e.g., in scientific papers, would have to conduct further systematic and in-depth investigations.

AI Learns Atari Arcade Games

Since 2014, a Google DeepMind team has been generating buzz with some spectacular presentations in which a machine learning algorithm self-learned classic arcade games without any prior human knowledge just from the feedback of the pixel images and game score and was able to improve to a superhuman level in a matter of hours (Hassabis 2014).

The developers' goal was to develop an algorithm capable of automatically developing a wide range of skills for a variety of challenging tasks—a central goal of general artificial intelligence. Although reinforcement learning had already achieved some success in a variety of domains, its applicability had until then been limited to domains that had fully observable and low-dimensional state spaces or to scenarios in which the artificial agent could be manually "handed" a set of useful features to help it deal with a complex environment.

In Kavukcuoglu et al. (2015), a corresponding deep Q-Network was developed by the developers. The DQN allows to combine reinforcement learning with deep convolutional networks ▶ (Sect. 5.1.3). In these networks, numerous layers are used to build progressively more abstract representations of the data ▶ (Sect. 5.1.4). Recurring simple basic building blocks ("features") are recognized from raw sensory data and can be synthesized into more complex representations.

The networks can therefore preprocess and simplify the state space to be handled from the high-dimensional sensory inputs, namely, 2D representations of arcade games, which allows generalizing similar sensory states and the corresponding "experiences" of the agent.

If we use these capabilities of CNNs to approximate and combine with Q-learning, then we need to take into account that reinforcement learning can be unstable for nonlinear function approximators. The instabilities described above when using deep networks to estimate the action value function $Q^*(s_t, a_t)$ was countered in the DQN from [9] with two measures: first, "experience replay," which eliminates unwanted correlations in the observation sequence by randomizing the order of the training data, and second, an iterative update, which adjusts only recurrently updated action values (Q) to prevent false links to the target state.

The algorithm proceeds roughly as follows: The action value estimator $\hat{Q}(s,a,\theta_i)$ is parameterized with the help of a deep-CNN, where with θ_i denotes the weights of the Q-network in the respective iteration i: The input to the neural network consists of an 84x84x4 image from direct preprocessing, followed by three convolutional layers and two fully connected layers with a single output for each valid action. Negative results from the hidden layers are $y = max\ (0,x)$ "truncated." For the "experience replay," the agent's experience is $e_t = (s_t, a_t, r_t, s_{t+1})$ at each time step t are stored in a data set $D_t = (e_1, e_2, ..., e_t)$ stored. Now, during learning, the Q-learning updates are applied on a random sample set that is uniformly randomly sampled from the pool of stored samples.

Furthermore, the target states in this setting also depend on the network weights, which is different from scenarios where these states are fixed before learning begins. The Q-learning update at iteration i uses the descent function:

$$L_i(\theta_i) = E_{(s,a,r,s')} \sim U(D)\left[\left(r + \gamma \max_{a'} Q(s',a'|\theta_i^-) - Q(s,a|\theta_i)\right)^2\right]$$

Here θ_i the parameters of the Q-network in iteration i and θ_i^- are the network parameters used to compute a target state in iteration i. The target network weights θ_i^- are kept fixed between updates. The original papers available online contain a comprehensive and clear presentation of the methods and technical details. It is hoped that the presentations in this book could at least help readers to access such papers—which, by the way, are often also freely available.

5.3   Neural Estimators for Action Selection

5.3.1   Policy Gradient with Neural Networks

As we have already seen, it is not necessary to consider explicit evaluations of environmental states in order to generate purposeful behavior. In simple reflexive biological behaviors, sensory impulses are conducted fairly directly to motor cells via a network of nerves. This can be observed, for example, in the nervous systems of simple creatures such as the marine snail Aplysia (cf. [8]). Here, "learning" takes place by reinforcing or inhibiting those connection parameters that produced a favorable or unpleasant result (or were associated with the event by the learning system).

Can we envisage neural networks that convert sensory signals into action probabilities and are adapted according to "experiences"? In principle, this corresponds to the logic of tactic search as we have investigated it in ▶ Sect. 4.2 for simple gridworlds.

In practice, different architectures and parameter values prove themselves. The obvious solution is to let the action references $h(s, a)$ learned by a neural network:

$$\pi\left(s,a\right)=\frac{e^{h_{net}\left(s,a\right)}}{\sum_{b\in A}^{A}e^{h_{net}\left(s,b\right)}}$$

However, in the actor-critic example presented below, these values are not learned directly, but the linear feature parameters $\boldsymbol{\theta}$ which are used to calculate the action references $h(s,a,\boldsymbol{\theta})$ are estimated with the help of the output of an MLP. $\boldsymbol{\theta} = \boldsymbol{net}\theta[\boldsymbol{x}(s,a)]$:

$$h_{net}\left(s,a\right)=\boldsymbol{net}_{q}\left[\boldsymbol{x}\left(s,a\right)\right]^{T}\cdot\boldsymbol{x}\left(s,a\right) \qquad (5.9)$$

Finally, as usual, the probability distributions of the policy are calculated from the action references using a softmax function.

The Java implementation is quite similar to that in Sect. 4.2.3 for the simple AC. It is only now that behind the functions that contain the evaluations V or the policy parameters $\boldsymbol{\theta}$ are not "tabular" structures, such as lists and maps, but neural networks that estimate these values by means of the state or feature vectors passed. The output error (loss) for the application of backpropagation results with:

$$\Delta\boldsymbol{net}_{q}\left[\boldsymbol{x}\left(s,a\right)\right]=h_{q}\,I\,d\nabla\ln p\left(s\right)$$

where δ the simple TD error is:

$$\delta=r+\gamma\hat{V}\left(s'\right)-\hat{V}\left(s\right)$$

$\hat{V}$ is also estimated with a neural network. The policy gradient is calculated with

$$\nabla\ln\pi\left(s,|a,|\boldsymbol{net}_{\theta}\right)=\boldsymbol{x}\left(s,a\right)-\sum_{b}\pi\left(s,|b,|\boldsymbol{net}_{\theta}\left[\boldsymbol{x}\left(s,b\right)\right]\right)\boldsymbol{x}\left(s,b\right)$$

computed. The training data is collected in small batches in random order and trained at regular intervals to stabilize the learning process somewhat.

Update of an Actor-Critic with ANNs for State Evaluation V and the Parameter Vector θ, That Weights the State Features for the Agent Policy in Java

```
protected void update( State  s,  int  a,  double  reward,  State
s_new,
                                                        bool-
ean episodeEnd ){
double[] x_s = s.getFeatureVector();
double observation = 0.0; if (episodeEnd) {
      observation = reward;
  }else{
            observation  =  reward  +  (GAMMA  *  getV(s_new.
getFeatureVector()));
  }
```

```
// TD-error ("1-step advantage")
double V_is = getV(x_s);
double delta=observation-V_is;
// Update "critic"
addV_MLP(x_s,observation);
// Update "actor"
double[] gradient = gradient_ln_pi(x_s,a);
double[] theta = p_theta[a];
for (int k=0;k<theta.length;k++){
    p_theta[a][k] += ETA_theta*I_gamma*delta*gradient[k];
}
addTheta_MLP(get_x_sa(x_s,a),theta);
I_gamma = GAMMA*I_gamma;
// batch update
cnt_updates++;
if ((cnt_updates%mini_batch_size==0)){
    network_theta.learn(trainingSet_theta);
    trainingSet_theta.clear();
    network_V.learn(trainingSet_V);
    trainingSet_V.clear();
    cnt_updates=0;
}
}
```

◉ Figure 5.21 shows the learning progress with an Actor-Critic Lunar Lander. The feature vectors were generated in the same way as for the Sarsa Lunar Lander. In addition to the network that approximates the weighting function $V(s)$ approximates the evaluation function, a network is now provided which approximates the policy parameters θ for a feature vector $x(s,a)$. The MLP for the evaluation function $V(s)$ has 18 neurons in the input layer in the 55 hidden layers and 1 neuron in the output layer. The MLP for action reference production θ has neurons; 23 in all 3 layers, since for each element of the feature (dimension: 3 by 7) action vector (dimension: 2) (cf. ▶ section "Feature vectors and partially observable environments"), an input and an output unit is provided. Each binary "feature" is assigned a parameter by the network, which determines the weight in the calculation of the respective action reference $h(s,a,\theta_S) = \theta_s^T x(s,a)$. Since we form the action references by a linear product of the parameter and the feature vector, the policy gradient here is the same as described in ▶ section "Feature vectors and partially observable environments".

5.3.2 Proximal Policy Optimization

Some methods that have only recently become known draw attention away from the policy gradient and look at the rapid accumulation of "positive and negative

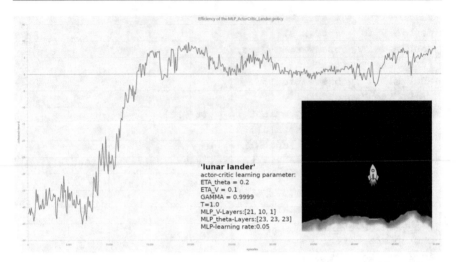

Fig. 5.21 Learning progress for the "LunarLander" with actor-critic with MLPs for V(s) and for the policy parameters θ. Each three sensors with seven binary features were shown. (Learning parameter: $\eta\theta = 0.2$; $\eta V = 0.1$; $\gamma = 0.9999$; MLP-V layers: [21, 10, 1]; MLP-θ layers: [23, 23, 23] The results of 100 episodes each were averaged

surprises" ("advantages") when determining action preferences. The "trusted region" methods show amazing performance while maintaining relative simplicity. The problem with this approach is that the policy must not be adjusted too quickly, i.e., the old and new policies must not diverge too much. We therefore need a learning rule that, in a certain sense, solves the trade-off between openness and conservatism as optimally as possible, i.e., on the one hand, learns quickly from the observations but at the same time does not discard the tried and tested too lightly.

In supervised learning methods, learning progress is generated by processing the difference between the output produced by the system and the "desired" output, called the "loss." In the policy gradient methods, the "loss" is calculated by:

$$L^{PG}(\theta) = \hat{E}_t\left[\log \pi_\theta(a_t|s_t)\hat{A}_t\right]$$

where $\hat{A}_t$ is an estimate of the Advantage function at time step t, e.g., formed by an n-step observation. For n = 1 $\hat{A}_t$ has the known value of simple temporal difference.

First, the ratio of the values of the new and old policy plays an important role. This probability ratio:

$$r(\theta) = \frac{\pi_\theta(a_t|s_t)}{\pi_{\theta_{old}}(a_t|s_t)}$$

indicates how much a currently present policy already deviates from an old one. If $\theta = \theta_{old}$,then the ratio is equal to 1. In Schulman et al., (2017) referring to a 2002 paper by Kakade and Langford,[4] present a loss that uses this likelihood ratio to adjust network parameters:

$$L^{CPI}(\theta) = \hat{E}_t \left[\frac{\pi_\theta(a_t|s_t)}{\pi_{\theta_{old}}(a_t|s_t)} \hat{A}_t \right] = \hat{E}_t \left[r(\theta)\hat{A}_t \right]$$

where CPI stands for "conservative policy iteration." Without constraints, the adjustment over the L^{CPI}, however, leads to excessively large policy changes. In the case of trusted region policy optimization (TRPO; Schulman et al. 2015), it is proposed to introduce a "penalty term" to prevent too large adjustments. For this, the "Kullback-Leibler divergence" $KL(P_1, P_2)$ which is a measure of how much two probability distributions differ. TRPO also aims to L^{CPI} maximize the KL divergence, but under the condition that the distance between the old and the new policy, as measured by the KL divergence, is small:

$$\underset{q}{maximize}\, \hat{E}_t \left[\frac{p_q(a_t|s_t)}{p_{q_{old}}(a_t|s_t)} \hat{A}_t \right] \text{provided that } \hat{E}_t \left[KL\left[p_{q_{old}}(\cdot|s_t), p_q(\cdot|s_t) \right] \right] \leq d$$

TRPO is relatively complicated to implement in practice A method based on the theory of TRPO, "proximal policy optimization" (PPO; Schulman et al. 2017), shows similar performance but using a surprisingly simple clipping function:

$$L^{CLIP}(\theta) = \hat{E}_t \left[\min(r_t(\theta)\hat{A}_t, clip(r(\theta), 1-\epsilon, 1+\epsilon)\hat{A}_t \right]$$

The first term within the min function of the "conservative policy iteration" loss, the second term, the one with the "clip" function, modifies the target by trimming the probability ratio so that it is kept within the interval $[1-\epsilon, 1+\epsilon]$ is kept. In this scheme, the pruning of the probability ratio is done when the objective function worsens or ignored when it would improve.

Proximal Policy Optimization, Actor-Critic Style (Hebb 1949)

```
1     for iteration=1, 2, ... do
2          for actor=1, 2, ... ,N do
3               Run policy π_old in environment for T timesteps
4               Compute advantage estimates Â_1, ... , Â_T
5          end for
6          Optimize L^CLIP wrt θ, with K epochs and minibatch size M ≤ N·T
7          θ_old ← θ
8     end for
```

---

[4] S. Kakade and J. Langford. "Approximately optimal approximate reinforcement learning". In: ICML. Vol. 2. 2002, pp. 267–274.

The algorithm of the PPO is designed for use in an implementation with parallel-creating working threads. However, you can also get an impression in our Greenfoot simulation in a single-thread implementation by overloading the update method of the actor-critic.

PPO Update in Java

```
protected void update( State s, int a, double reward, State s_new,
boolean
episodeEnd ){
        double[] x_s = s.getFeatureVector();
        double observation = 0.0;
        if (episodeEnd) {
                observation = reward;
        } else {
                observation = reward + (GAMMA * getV(s_new));
        }
        // TD-error ( '1-step Advantage' )
        double V_is = getV(s);
        double delta=observation-V_is;
        // Update critic
        addV_MLP(x_s,observation); // add record to net_V minibatch
        // Update actor
        double[] pi_s = P_Policy(s);
        advantages.add(new AdvPi(s,a,delta,pi_s[a]));
        if (cnt_steps%horizon==0) {
                while (!advantages.isEmpty()){
                        AdvPi e = advantages.removeLast();
                        double[] pi = P_Policy(e.s);
                        if (!determ_pi(pi_s)){
                                double r_theta = (pi[e.a]/e.pi_sa);
                                 double delta_theta = ETA_ppo*min(r_
theta*e.adv,
                                 clip(r_theta,1-EPSILON_ppo,1+EPSILON_
ppo)*e.adv);
                                double[] exs = e.s.getFeatureVector();
                                double[] theta = get_h(exs);
                                h[e.a]= h[e.a]+delta_theta;
                                addH_MLP(exs,theta); //record to net_h
minibatch
                        }
                }
        }
}
```

For states with quasi-deterministic decisions, infinities can occur in the action references, which should be caught by the determ_pi function; if the probability on an action is very close to 1.0, the function returns true and prevents further matching and a possible overflow.

With η_{PPO} a kind of step size was also added. A value of $\eta_{PPO} = 0.5$ sometimes showed a more stable learning progress. With appropriately selected parameters, which is not always easy, PPO shows exceptional performance even in our "game environment" and with only one thread. In individual test runs with the "LunarLander," the use of RBF features proved favorable (cf. ◉ Fig. 5.16).

Technically, the next challenge is now to implement A3C and PPO with neural networks and threads working in parallel. This is primarily a software engineering challenge. You can find corresponding implementations in the companion materials to the book for download. For professional applications and learning speeds, the algorithms are connected to an appropriate machine learning engine, which can also optimally use current hardware for training the neural networks. For further details on "deep reinforcement learning," please refer to the relevant literature. For more in-depth information on the theory and practice of deep learning with Java, you may also find suitable information on the page (https://d2l.djl.ai/index.html) (Fig. 5.22).

In the last technical section, an evolutionary strategy with neural networks will be presented once again.

5.3.3   Evolutionary Strategy with a Neural Policy

Artificial neural networks are very well suited for the implementation of an evolutionary strategy. For the use of neural networks in a genetic algorithm, the structure of the network, i.e., at least the connection weights, must be mapped in the genome.

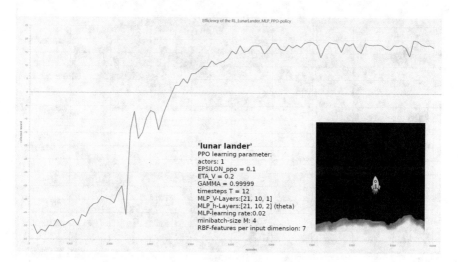

Fig. 5.22 Learning progress in the LunarLander with proximal policy optimization (PPO). RBF features were used (seven per input dimension). The results of 100 episodes each were averaged

Adaptation then takes place, as before, by carrying out random modifications ("mutations") to the genome and screening out and propagating the agents that work best.

An example of this can be found in the scenario "Evolutionary RoboCarts." This is a simulation of two-wheeled robot cars with differential drive, similar to the "Lego" or "Makeblock" robots presented at the beginning (see Fig. 5.23).

The neural network of a robot (object "brain") is generated from the genome in the constructor of the robot.

Constructor of an Individual Robot

```
public    AdaptiveMbRobot(MbRobotSensors    sensors,    MbRobotMotors
motors, Genome
gene ){
 super(sensors,motors);
 this.ID = AdaptiveMbRobot.lastID;
 AdaptiveMbRobot.lastID++;
 this.crashed=false;
 this.sum_reward = 0.0;
 this.battery_charge=MAX_BATTERY_CHARGE;
 this.gene = gene;
 this.brain = new NeuralNetwork(NEURAL_NET_STRUCTURE);
 this.brain.createWeightsFromGenome(gene);
}
```

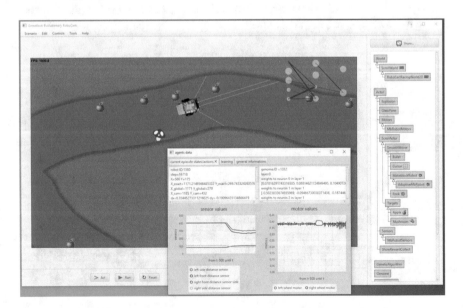

Fig. 5.23 Simulation of a two-wheeled robot with genetic algorithm

The neural network in the scenario is not implemented with Neuroph or another framework but was coded "manually." Without learning rules like "backpropagation", only the corresponding matrix multiplications have to be performed for each layer by multiplying the respective input activation with the connection weights.

Generation of a Neural Network from a Genome

```
public void createWeightsFromGenome(Genome genome) { if (genome
== null)
return; network_weights = new ArrayList<double[][]>(); int coun-
ter=0; for
(int layer=1; layer<size_of_layer.length; layer++){
        double[][] weights =
                        new   double[size_of_layer[layer]]
[size_of_layer[layer-1]+1];
                                                            // +1
because of BIAS
        for (int m=0;m<weights.length;m++){ // outputs "to neuron
                for (int n=0;n<weights[0].length;n++){ // inputs
"from neuron
                        weights[m][n]=genome.weights.get(counter);
                        counter++;
                }
        }
        network_weights.add(weights.clone());
    }
}
```

In the present example, however, continuous motor values are produced rather than action references being mapped by the genome as in Chap. 4. Thus, this time we are not dealing with a stochastic policy, but with a deterministic policy that converts continuous sensor values into continuous action values, i.e., $s \in S \subset R^n$; $a \in A \subset R^n$.

For the understanding of the evolutionary algorithms, the environment class that generates and sifts the populations is very important. In the presented simulation, this is done by the RoboCartRacingWorld2D class. After starting the simulation, a population of robots is created on a track. The initially chaotic behavior becomes increasingly orderly. The robots receive rewards by collecting apples. This also recharges their energy.

The scenario contains many technical details, such as the scroll mechanism, the sensor beams, or the possibility to select individual robots. In addition, not only the fitness progress but also the sensory inputs and motor outputs for the selected robot can be followed "live" in the Javafx logging monitor. The neural network of the selected robot is also visualized.

However, you should not be put off by the complexity of the software. The act function of the agent is very simple. With genetic methods, we do not need any learning processes within the agent, i.e., no individual update, unless we also optimize the learning processes themselves with an evolutionary strategy.

Act Method in "Evolutionary RoboCarts"

```
public void act() {
  s = getState();
  a = policy(s);
  setAction(a);
  move();
  sum_reward += getReward();
  this.gene.fitness = sum_reward;
  if (this.isCrashed()) world.removeCrashed(this);
  if (connectedCursor!=null) world.updateStateCurveMonitors(this,
cnt_steps,s,a);
  cnt_steps++;
}
```

In principle, the learning process follows the procedure presented in Chap. 4 of selecting and propagating those individuals with the greatest fitness after the completion of an episode (all robots are crashed or stalled). The makeNextGeneration() function in the RoboCartRacingWorld2D class is called when it is determined that a new episode needs to be started. This happens here when no individual of the population is alive anymore.

Creation of the Next Generation from the Selected, Crossed, Recombined and Mutated Genomes

```
public void makeNextGeneration(int globalX, int globalY){
  robotPopulation = new
                     AdaptiveMbRobot[geneticAlgorithm.getGenePool().
size()];
  geneticAlgorithm.breedPopulation();
  ArrayList   <Genome>   currGenePool   =   geneticAlgorithm.
getGenePool();
  int i=0;
  for (Genome gene : currGenePool){
      robotPopulation[i] = produceRobot(gene);
      robotPopulation[i].setCrashed(false);
      addObject(robotPopulation[i],0,0);
      robotPopulation[i].setGlobalLocation(globalX, globalY);
      i++;
  }
  cnt_crashed=0;
  if (jfxMonitor!=null) jfxMonitor.clearEpsiodeData();
}
```

It is fascinating to observe how some robots complete the course well after only a few generations. The general setting is very informative and also again vividly illustrates the roots of reinforcement learning in the evolution of purposeful behavior of "situated" systems that must be successful in their environment.

Bibliography

Alpaydin E (2019) Machine learning, 2nd expanded edition (De Gruyter Studium)

Been K, Pavlus J (2019) A new approach to understanding how machines think. Quantamagazine. Available online at: https://www.quantamagazine.org/been-kim-is-building-a-translator-for-artificial-intelligence-20190110/

Churchland PS, Sejnowski TJ (1997) Fundamentals of neuroinformatics and neurobiology. The computational brain in German: vieweg computational intelligence

Frochte J (2019) Machine learning: foundations and algorithms in Python, 2nd edn. Hanser Verlag, Munich

Fyfe C (2007) Hebbian learning and negative feedback networks, Advanced information and knowledge processing. Springer, Dordrecht. Available online at: http://gbv.eblib.com/patron/FullRecord.aspx?p=371973

Hassabis D (2014) Deepmind artificial intelligence @ FDOT14. Available online at: https://www.youtube.com/watch?v=EfGD2qveGdQ

Hebb DO (1949) Organization of behavior

Kandel ER (2009) In search of memory. The emergence of a new science of the mind. Paperback edition, 4th edn. Goldmann, Munich (Goldmann, 15570)

Kavukcuoglu K, Minh V, David (2015) Human-level control through deep reinforcement learning. Nature. Available online at: https://web.stanford.edu/class/psych209/Readings/MnihEtAlHassibis15NatureControlDeepRL.pdf

Ribeiro MT, Singh S, Guestrin C (2016) "Why should I trust you?" Explaining the predictions of any classifier. In: Proceedings of the 22nd ACM SIGKDD international conference on knowledge discovery and data mining, pp 1135–1144. Available online at: https://arxiv.org/abs/1602.04938

Schulman J, Levine S, Abbeel P, Jordan M, Moritz P (2015) Trust region policy optimization. In: Proceedings of the 32nd international conference on machine learning, PMLR 37, pp 1889–1897

Schulman J, Wolski F, Dhariwal P, Radford A, Klimov O (2017) Proximal policy optimization algorithms. Preprint at https://arxiv.org/abs/1707.06347v2

Sutton RS, Barto A (2018) Reinforcement learning. An introduction, 2nd edn. The MIT Press (Adaptive computation and machine learning), Cambridge, MA/London

Turing A (1937) On computable numbers, with an application to the decision problem. In: Proceedings of the London mathematical society, vol 42, ISSN 0024-6115, pp 230–265, Available online at: https://londmathsoc.onlinelibrary.wiley.com/doi/abs/10.1112/plms/s2-42.1.230 (Oxford Journals)

Guiding Ideas in Artificial Intelligence over Time

6

Abstract

In this chapter, we will take a somewhat philosophical look at the changes in the guiding ideas of artificial intelligence research and how they have changed over time. Finally, we will take a look at the relationship between humans and artificial intelligence and how to make sense of it.

6.1   Changing Guiding Ideas

An early seminal classification of systems behavior was published by Arturo Rosenblueth, Norbert Wiener, and Julian Bigelow as early as 1943 in an article in the journal Philosophy of Science (Rosenblueth et al. 1943). The publication of the three renowned scientists was likely to stimulate a wave of discussion among researchers in many disciplines, including philosophers, biologists, neurologists, and later among scientists in the emerging field of computer technology. A distinctive feature of the article was that it described certain forms of system behavior as "intentional" and thus "purposive" from the perspective of mechanism. In doing so, they created a classification of behavior:

1. Active
2. Purposeful
3. With "feedback" (teleological)
4. Predictive ("extrapolative")
5. With higher levels of prediction ("first-, second-, etc. orders of prediction")

They describe system activities as "purposeful" or "intentional" if they do not occur randomly but are directed toward a specific goal. It was remarkable that attributes such as "purposeful" and "intentional" appear as objectively measurable properties of a system. Action choices deviate from chance in a well-founded way. The reasons, i.e., the objectives of a purposeful acting system depend on the environment, the "world context," but no less on the task a system possesses, i.e., on its purpose or "system context." The intrinsic reasons for activity, in turn, derive from external processes and are to be seen in connection with the causes and reasons for the formation of the system. The task fundamentally determines the way in which the system engages with the environment. The system structure must correspond to the "interests" and "tasks" of a system, it must ensure the useful transformation of environmental influences.

This "cybernetic" view was temporarily pushed into the background by the emergence of digital technology. A view became dominant which today is often referred to by the keyword "GOFAI." In this approach, "objectives" are conceived as logical "problems," as in the physical symbol systems hypothesis (PSSH) put forward by Allen Newell and Herbert A. Simon in 1975. The PSSH states that a physical symbol processing system provides "in principle the necessary and sufficient conditions for intelligent action"[1]. The traditional GOFAI approach added to this hypothesis the basic assumption that mental processes are, in principle, themselves processes of logical symbol processing. In the traditional symbol processing paradigm of A.I., it is assumed that symbols and symbol structures can be linked to specific elements or processes from the system's environment. Thus, the symbol structures in the computer system's memory are supposed to stand for something in the external world in order to model this environment inside the machine. The symbols are thereby understood as "meaning carriers." On the basis of the environment model, which is supposed to represent reality, operations are carried out to calculate the best output option.

The purely "model-based" approach came under heavy pressure from critics from various directions. Rodney A. Brooks, for example, formulated a fundamental critique. In his article "Intelligence Without Reason" (Brooks 1991), he criticizes that the previous ideas were essentially influenced by the technological limitations of the available computer technology. He bases his arguments against the traditional approach on the one hand on results of biological behavioral research with theories

---

[1] From Gerhard Strube *Dictionary of Cognitive Science*.

in which concepts such as "motivational competition," "disinhibition," and "dominant and subdominant behavior" play a role. He argues that there are no explicit internal representations in biological systems and that purposeful behavior is produced by "complex internal and external feedback loops". Furthermore, he argues that it is completely unrealistic to create complete, objective models of the world. The experimental "successes" of the traditional approach are based exclusively on very carefully constructed, abstract "mini-worlds" ("block-worlds"). He also rejects approaches with "bounded rationality", e.g., by Russell, as an attempt to squeeze traditional AI with its fundamental problems into a system with finite computing capacity.

The unsolvable practical problems lead to the discussions on the so-called "frame problem." Where, in principle, the question is how an adequate and efficient environmental representation can be found. It turned out that "relevance" cannot be defined independently of concrete action. An example will illustrate this: If, for example, a car hits an apple tree, then not only the car is broken, but perhaps also the driver, the tree, etc., the birds have flown away, the apples have fallen down, etc., pp. How is one supposed to recognize and store all the correlations? Obviously this is not possible. Now how is a purely "knowledge-based" system supposed to distinguish between relevant and uninteresting information? The only complete "model" of the world would be the world itself, and this "model" is cheaper to have, so we would be back at the starting point.

Another argument underlines that intellectual activities in general, e.g., scientific research, are not only about the production of solutions to given problems, but also about the proper production of the problems themselves. The argument is relevant in relation to AI research on two levels: on the one hand, for the artificial systems with their own goals, which must be conceived as "agents" and have their own standard of evaluation—the keyword being the "reward" function—and, on the other hand, for the scientific study of artificial intelligence itself. The French philosopher Henrí Bergson (1859–1941) points out (Bergson 1991) that from the implicit problem and goal emerges the perspective and the "illumination," in principle our complete picture, of an object. If this "illumination" and the perspective is insufficient due to wrong problem or goal setting, then not only the object but also the solution remains dark. If, on the other hand, a scientific object is properly illuminated and the right questions are asked, then the solution will almost automatically emerge. This means that in basic research we are not only dealing with wrong solutions, but above all with wrongly posed problems. The GOFAI implication that intelligence in humans—and even more so in animals—essentially emerges from a symbolic model of the world was based on false premises and objectives.

However, it does not follow from the impossibility of modeling the world completely and effectively symbolically, i.e., in the sphere of "pure knowledge," that it is impossible to reproduce "cognitive apparatuses" by electronic, including digital, means. It is commonly assumed that the serial-algorithmic machine model developed in the tradition of mechanics and logic has only limited capabilities compared to, for example, the human brain. However, the opposite is actually the case: computer systems are much more general in their functioning than biological cognitive systems; they can not only control the actions of agents but also play movies or simulate atomic bomb explosions. They are more akin to a construction kit than to a cognitive system that has to cope with the tasks of everyday life. So it may not be entirely accurate for Brooks to suggest that the inadequacy of AI problems is due to the "limitations" of digital hardware. It is probably more accurate to say that the mistake was in transferring the computer's design principles, which were oriented toward formal calculus, to the level of producing intelligent behavior.

In model-free approaches, one refrains from constructing explicit environmental representations and focuses on the study of cognition, as "the processes intervening between perception and motor activity"[2]. The starting point is initially behaviorist. Of central interest is the appropriate relation of the system to its environment, rather than its "correct" internal organization. Accordingly, a cognitive system has a mediated interaction of perception and motor activity, whereby conditional actions are made possible. In "learning," the communication of these elements is adjusted until the system as a whole has the desired relationship between input and output.

The behavior-based conception of "learning" proved to be extraordinarily viable. Learning is now understood, exactly as we have conceived it in the context of "reinforcement learning," as a process by which a given system behavior changes in such a way that it corresponds ever better to certain quality criteria. In nature, these quality criteria have evolved phylogenetically and are usually oriented toward "viability and reproductivity." In artificial systems, they should be oriented to the task the system is supposed to fulfil for humans. Adaptation processes are now again feedback processes that essentially consist of two parts:

1. "Action": means anticipatory behavior based on given internal and external conditions.
2. "Criticism": constructive correction of the system through internal or external processes.

The task of the "critical" process is to reinforce "correct" parts and reduce "incorrect" ones. By the way, this also applies to "supervised learning". Such learning processes offer the possibility to avoid internal imitations of the external environment. The problem now is to construct the "direction of action" of the "critical moment" correctly, i.e., to implement the learning mechanism in such a way that the system body optimizes itself task-oriented by the feedbacks. The cognitive components, as activity-guiding links between the sensory and motor systems, do not initially have the task of producing an environmental model, but rather of generating appropriate system behavior.

In the biological field, all components of the "system body", i.e., the sensory, vegetative, and motor components, must constantly prove themselves with regard to the tasks of the living system. But if all parts of the system body have to prove themselves permanently in their usefulness, the body as a whole becomes an expression of the realized forms of behavior. Changes in behavior thus become changes in the

---

[2] Ibid.

system body at the same time. This leads us to the conclusion that there can be no universally valid presuppositions concerning the constitution of a system body that are independent of the system context. Since the tasks to be solved by machines or artificial agents may differ significantly from those faced by biological organisms, the corresponding machine "bodies" may also differ widely from what is understood by a body in the conventional biological sense. Why should a text recognition system or a chess computer have an "embodiment" in the biological sense? Only the corresponding input and output units, as well as the computational processing capacities, are needed for this.

Brooks responded to the crisis of GOFAI ideas with concepts such as embodiment and situatedness. "Embodiment refers to the importance of a body, without which an artificial system has no way of sensually interacting with and experiencing the world. For Brooks, an intelligent system is not primarily an intelligent "thinking" system, but an intelligent acting system. For him, true intelligence is inconceivable without situatedness. "Situatedness" means a permanent interactional relationship of a system with the environment in which it is embedded. However, his point here is not only to refer to this "embeddedness" but also to assert the uselessness of plans within an abstract, explicit model of the world. Although he raises important points and has stimulated much fruitful discussion, this apparently overshoots the mark, i.e., beyond what is useful for us.

When it is clear that cognitive processes and environmental models only serve competent behavior, then these "environmental models" can play a role again. Sutton and Barto, as key pioneers of reinforcement learning, show the way when they point out that the aim must be to leave this "ideological" argument behind and combine model-free with model-based approaches. It is exciting how, for example, in Dyna-Q they use the model to generate virtual observations. So the "model" here is used to generate "cheap" observations that optimize the model-free basis. Historically, model-free approaches initially offered themselves as more effective machine learning methods. This has changed. There are indications that model-based augmentations can serve to make the model-free methods more efficient, especially in more complex environments with few or "distant" target states.

Technical and scientific progress does not suddenly fall from the sky as a result of ingenious ideas, even if sometimes groundbreaking ideas produce leaps and bounds in progress. The first car still looked almost like a carriage. Progress comes from "pacing," that is, from "making," from hypothesizing, from building prototypes, from testing, adapting, observing, and improving. It may not have been right to start with a "knowledge-based" or "model-based" approach, but neither have model-free methods been the "last word in wisdom." At present, the picture is rather that policy-based, statevalue-oriented, i.e., model-free learning and model- or knowledge-based methods build on each other (cf. also ▶ Sect. 4.5), which also has a certain plausibility with regard to the sequence in the evolutionary history of natural cognition.

6.2   On the Relationship Between Humans and Artificial Intelligence

"In the logic of AI, there is no free will. Machines do what they have been programmed to do. They behave as they are supposed to." This was so stated by Julian Nida-Rümelin in an article in the Max Planck Society's science magazine "Max Planck Research" (issue 2/19). The article also introduces his current book "Digital Humanism," which he co-authored with Nathalie Weidenfeld. Perhaps the reader, even after reading this book, is also skeptical whether Mr. Nida-Rümelin is still correct with this statement.

> Computers aren't supposed to be creative; they're supposed to do what you tell them to.
> If what you tell them to do is be creative, you get machine learning. (Domings 2015)

The Google Translator currently (2019) translates this sentence by Petro Domings into German like this: "Computer sollen nicht kreativ sein; sie sollen das tun, was man ihnen sagt. Wenn man ihnen sagt, dass sie kreativ sein sollen, entsteht maschinelles Lernen." No programmer programmed into the Google Translator how to translate that sentence. Neural networks play a role in such translators, as they do in the deepL translator, and they continually improve themselves.

To understand the computer only as a calculating automaton, which manipulates symbols and thereby mechanically comes to more or less meaningful results, depending on how well the rules were defined, according to which this manipulation takes place, certainly no longer corresponds to the state of development. Describing modern software systems in terms of mechanical computing is similar to trying to describe, for example, an earthworm with the atomic model, if only one considers that computers in the range of two-digit TFlops (= thousand billion floating point operations) are available for private users today and at the same time several billion bits of working memory are standard.

Of course, a new technology such as machine learning also poses dangers that can only be limited by responsible and competent use of the technology and appropriate legal rules. In Rümelin's critique quoted above, however, the dangers posed by the new technology are not the focus. Rather, he sees classical humanism in danger because of the naturalistic, or materialistic, approach of AI research and takes a defensive stance. "Digital humanism does not transform humans into machines and does not interpret machines as humans" the article states. "It holds to the distinctiveness of humans and their capabilities, and uses digital technologies to augment them, not limit them."

The "exploration of mental processes," as in the quotation from Burkhard presented at the beginning, is, however, a profoundly humanistic aspiration, which, incidentally, goes hand in hand with the fundamental questions of philosophy. Up to now, one could only devote oneself to this task by thinking and discussing, thus one had to rely solely on the use of one's brain. With modern digital technology and its possibilities of simulation, we have a new kind of instrument at our disposal for this research work, which provides us with exciting new insights and possibilities. Viewed from this vantage point, the naturalistic view precisely enables us to "make use of digital technologies [in order] to extend these [human capacities and characteristics], not to limit them." Philosophy is actually extended by this technology into an empirical realm. It becomes possible to look at philosophical insights or questions, e.g., related to the relationship between consciousness, mind and matter, subject and object, freedom and determinism, symbol and concept, etc., from a new perspective and also to investigate them experimentally. Is not the very view limiting and mythological that for us the principles of intelligence and thought must ultimately remain incomprehensible?

From an anthropological point of view, the large brain with its abilities for language, instrumental intelligence, empathy, etc., with its possibilities to recognize the goals and reasons for action of others and to share these supra-individually, is certainly a very characteristic feature of man, perhaps like the oversized beak of the stork or the nail tooth of the beaver.

Nevertheless, our "being special" is not challenged by the invention of "intelligent" machines. Working with handicapped people at an inclusive education center, with highly gifted teenagers up to severely damaged children and adolescents, the author of these lines could see that "intelligence" does not play any role at all for the definition of "humanity."

That which for us represents the "specialness of man" is in principle just as little called into question by a partial overtaking of our biological abilities by machines as it has happened by the invention of the excavator, although this in a certain way far surpasses our abilities. Furthermore, perhaps it is reassuring to think that even a hypothetical superhumanly complex A.I., because of its different "embodiment" and radically different history of creation, would always be denied the ability to really know what it "feels like" to be human. Just as little as we can understand how it "feels" to be, for example, a honeybee.

The state of the art in reinforcement learning is such that we can not only learn a lot of important things from the functioning of the new autonomously acting intelligent agents, but that it can also be very exciting to examine the results produced by these "intelligent" algorithms. Garry Kasparov (former world chess champion) commented on the performance of the agent AlphaZero developed by Google DeepMind:

> I'm amazed at what can be learned from AlphaZero, and basically from AI programs that can recognize rules and paths previously hidden from humans... The implications are obviously wonderful and far beyond chess and other games. The ability of a machine to copy and outperform centuries of human knowledge in a complex, closed system is a tool that will change the world.

It may soon be that a larger proportion of scientists will be involved in analyzing the results produced by artificial intelligences. The systems do not represent "black boxes" either, but give humans a complete insight into the processes taking place without any resistance. The problem is to see through the complexity and to discover the rules according to which the machine has generated the "competences." There is already a great deal of research into how the processes of machine learning can be made comprehensible and transparent; the keyword here is "explainable arti-ficial intelligence," or "XAI" for short.

Some fear that A.I. will make man a "discontinued model." Since "man" is the designer of the development of technology, this would mean that man "disposes" of himself with AI. Ideas that do things that are commonly unintended should be iden-tified as "wrong" because they are not only pointless, but harmful. However, - why do people, whether "up to date" or "phased out," actually consider themselves a "model"? There is indeed an equation of man and machine that is worthy of notice and criticism.

To resolve this equation, however, we do not need to eliminate AI. As purpose-lessly created beings, we don't have to and shouldn't "function." It is fantastic that we can create possibilities to transfer "functional tasks"—i.e., tasks that transform "inputs" into "target outputs"—to machines. To be sure, this would indeed make the biological "machine man" obsolete. The "real human being", however, if he wants to be happy, has to cope with tasks of a completely different kind than those we set our machines.

What do we want and should we transfer to machines at all, and what do we prefer to keep in our own hands? Furthermore, artificial intelligent systems should never be "sure" that they have correctly understood the task and the motive of humans, as Stuart Russell emphasizes. Machine systems that are capable of learning must make their decision-making transparent to us, and, if necessary, let us have the last word. The contradictions and problems of human societies will necessarily be reflected in them. It is clear that a number of things must be taken into account when handing over tasks to machines. Many smart people have been thinking about this for a long time. Learning systems are shaped by their environmental system and act back on it. The central pivot of the environmental system of more complex AI's must be the human being with his needs. This can also include, for example, their need to be "left alone." Unfortunately, it is to be feared that in our present state of society also some terrible nonsense will be done with these tools.

It is an old discussion: Should a technological development be stopped if compa-nies and states use this technology in a way that affects people—for example, through unemployment or wars? It might be more sensible and more promising to first question the foundations of the negative developments, e.g., profit orientation and competition. Otherwise, these problems will always be created anew. If, for example, search engines or "social" networks exploit the data of users in such a way that they are manipulated or incapacitated, would it not then make more sense to operate this technical infrastructure, for example, through nonprofit foundations or public institutions, according to transparent rules and democratically influenceable objectives? Such causal changes would also bring about a different development of technology—perhaps one that actually enables us to have the hoped-for all-round global exchange, new exciting insights, highly expedient processing of information of all kinds, and the elimination of involuntary, burdensome labor.

Bibliography

Bergson H (1991) Matter and memory. A treatise on the relation between body and mind. Meiner, Hamburg (Philosophische Bibliothek, 441)

Brooks RA. Intelligence without reason. In: IJCAI'91, pp 569–595

Domings P (2015) The master algorithm. How the quest for the ultimate learning machine will remake our world. Penguin Books, London

Rosenblueth A, Wiener N, Bigelow J (1943) Behavior, purpose and teleology. Philos Sci:18–24